SEMICONDUCTORS
Integrated Circuit Design
for Manufacturability

SEMICONDUCTORS
Integrated Circuit Design
for Manufacturability

Artur Balasinski

CRC Press
Taylor & Francis Group
Boca Raton London New York

CRC Press is an imprint of the
Taylor & Francis Group, an **informa** business

CRC Press
Taylor & Francis Group
6000 Broken Sound Parkway NW, Suite 300
Boca Raton, FL 33487-2742

First issued in paperback 2017

© 2012 by Taylor & Francis Group, LLC
CRC Press is an imprint of Taylor & Francis Group, an Informa business

No claim to original U.S. Government works

ISBN-13: 978-1-4398-1714-8 (hbk)
ISBN-13: 978-1-138-07541-2 (pbk)

Library of Congress Cataloging-in-Publication Data

Balasinski, Artur.
    Semiconductors : integrated circuit design for manufacturability / Artur Balasinski.
        p. cm. -- (Devices, circuits, and systems)
        Summary: "Preface The discipline of Integrated Circuit Manufacturing (ICM) and the field of Design-for-Manufacturability DfM related to it, are extremely dynamic fields in technical concepts and market applications. This dynamism makes writing subject matter publications both easy and difficult. It is easy enough to provide an up-to date snapshot of ICM DfM issues, given the breadth of disciplines and the influx of publications. But is it harder to ensure any longevity of the existing technical solutions. Technical knowledge is often dated when already in the press. Therefore, the concept of this book is to focus on trends and correlations with which to define ICM DfM. DfM approaches proposed in other engineering fields and proven wrong 100 years, 50 years, or 1 year ago, have a high probability of being wrong in the future. A classic example is downplaying reliability to the short-term advantage of marketability. On the other hand, approaches successful in the past, such as correct-by-construction (CBC), are likely to positively impact product development also in the years to come"-- Provided by publisher.
        Includes bibliographical references.
        ISBN 978-1-4398-1714-8 (hardback)
        1. Integrated circuits--Design and construction. 2. Integrated circuits--Design and construction--Cost control. 3. Semiconductor industry--Cost control. I. Title.

TK7874.B35 2011
621.3815'3--dc22   2011010690

Visit the Taylor & Francis Web site at
http://www.taylorandfrancis.com

and the CRC Press Web site at
http://www.crcpress.com

To my son, Chris

When you design your life,

first you have choices, but no

commitments; then commitments,

but no choices. Make it simple:

Commit to a good choice.

Contents

Preface

The disciplines of integrated circuit manufacturing (ICM) and of design for manufacturability (DfM) related to it are extremely dynamic fields in technical concepts and market applications. This dynamism makes writing subject matter publications and monographs both easy and difficult. It is easy enough to provide an up-to-date snapshot of ICM DfM issues given the breadth of disciplines and the influx of publications. But it is harder to ensure any longevity of the existing technical solutions. Technical knowledge is often already dated when presented in the press. Therefore, the focus of this book is on trends and correlations with which to define ICM DfM. DfM approaches proposed in other engineering fields and proven wrong 100 years, 50 years, or even 1 year ago have a high probability of being wrong in the future. A classic example is downplaying reliability to the short-term advantage of marketability. On the other hand, approaches successful in the past, such as correct by construction (CBC), are likely to have a positive impact on development of any product in any manufacturing discipline in the years to come.

Acknowledgments

I thank the many friends and colleagues of Cypress Semiconductor for inspiration and the opportunity to work on this publication. Special thanks go to Mrs. Agnieszka Baranowska for editorial assistance.

Artur Balasinski

About the Author

Artur Balasinski received the PhDEE degree in MOS (metal oxide semi-conductor) technology from Warsaw University of Technology, Warsaw, Poland in 1987, where he continued as assistant professor. He then joined the research team at Yale University, New Haven, Connecticut to continue studies on rad-hard devices. Subsequently, he entered the integrated circuit (IC) industry, first in research and development at STMicroelectronics, working on CMOS process transfers, and, since 1997, at Cypress Semiconductor, where, as principal technology-design integration (TDI) engineer, he developed expertise in characterization, process integration, optical proximity correction, and design rules. He has authored or coauthored about 90 papers (3 of them received best paper awards), a book chapter, and has 15 U.S. patents. A secretary of the Bay Area Chrome Users Society (BACUS) Photomask Steering Committee, he has, in recent years, attended SPIE (Society of Photographic Instrumentation Engineers or International Society for Optical Engineering), BACUS, VLSI (very large scale integration), and IEDM (International Electron Devices Meeting) conferences, where he formulated his views on DfM, presented several invited papers, and has held special sessions.

1

Introduction

1.1 Design for Manufacturability: Integrated Circuit Manufacturing versus Industry Legacy

Since the 1960s, integrated circuit manufacturing (ICM) has been making exponential progress, increasing the number of manufactured devices and the value of the total available market (TAM). The rate of this progress was such that it grew over a trillion times (12 orders of magnitude, 10^{12}) in that time period, with a potential to expand believed to be at least another 7 years onward from any given point in time [1]. In the process, ICM companies created multiple rules, guidelines, and best practices, known as design for manufacturability (DfM), to ensure efficient electrical and physical design and to expedite revenue growth. This methodology was elaborated on in thousands of papers, disclosures, and monographs. A large industry, electronic design automation (EDA), was brought to life and became indispensable to ICM for developing tools and solutions for DfM.

However, as an increasingly wider IC engineering audience realizes, ICM-DfM-EDA interactions are not seamless, and the continuous growth is putting strain on these seams. The IC DfM recital of problems and solutions has been created mostly in the reactive mode, oriented toward technical goals, which assume that IC (integrated circuit) development would for the predictable future follow the path of geometry reduction defined by Gordon Moore in 1962 [2]. While this assumption seemed to be justified by the longevity of the trend, it does not take advantage of concepts outside the immediate IC engineering area and does not elaborate on alternative scenarios. For example, for how long would the optical proximity correction (OPC) of IC layout, now considered a major part of DfM, be relevant if the IC industry makes the unavoidable switch from deep ultraviolet (UV) to one of the next-generation lithography technologies: nanoimprint or extreme ultraviolet lithography (EUV)? And that leap is only one of the many expected for the very large scale integrated (VLSI) manufacturing flow, which may soon divert (e.g., into building devices from organic materials). Would it mean the end of IC DfM as we know it?

Looking after its own problems, IC DfM stayed largely independent of DfM methodologies developed for other manufacturing disciplines with longer legacies. For example, mechanical engineering (ME) DfM methodologies go back 200 years and could have been referenced to support IC DfM. However, IC DfM and ME DfM developed independently of each other. As the current forces driving IC DfM reach maturity and ICM becomes a staple of engineering life, the incomplete links in this correlation may be about to converge. Both ICM and ME are ultimately about making profits in the marketplace, and both have to ensure success through aggressive cost reduction in addition to the relentless innovation. For IC DfM, it is becoming indispensable to follow such ME DfM concepts as concurrent engineering, cost commitment schedule, or linear thinking [3,4].

The IC growth defined by the exponential Moore's law since the 1960s was typical for young engineering disciplines and was not unlike its nineteenth-century predecessors. At that time, the transportation industry enjoyed similar, unprecedented growth; the shipbuilding tonnage exploded, followed by a similar increase in the automobile population and then by aircraft count and speed in the twentieth century. However, in all these cases, fundamental limitations related to the speed of travel, combined with market saturation, changed these industries from explosively growing to aggressively cost-cutting ones. DfM played a key role in both phases. In the first phase, design benefitted from increasingly more complex and accurate models for the upcoming product families, enabling manufacturing to build the products with confidence and yield, also requiring offshore, human labor that can be trained to compete successfully with that of the country of original invention. The ship- and car building moved from the cradle of manufacturing, England, to continental Europe and then to America and the Far East.

In the past, the high growth rate of TAM, number of devices, and applications made ICM impregnated in the conservative DfM approaches oriented at cost savings. Investment practices, instead of being frugal, relied on the momentum of continuous growth, knowing that the next generation of IC products would most likely bring in twice as much revenue as the previous one did, in line with Moore's law. In the process, IC DfM became reliant on improvements in layout conversion into silicon and overcoming pattern resolution problems to support the ever-shrinking dimensions of active devices.

The growth direction of IC DfM is now gradually evolving. Although technology still drives future development, the economy, which always has the final word, points at the need of fully exploiting the potential of existing applications before investing in new ones. The risk-reward shrink path of runaway IC geometrics-cost relationship would eventually lose steam at the level of a multibillion dollar investment. There seem to be fewer simple ideas to fuel the growth and fewer companies succeeding in generating returns at such cost. The revenue per unit area still has to increase (say, 40%) to allow for manufacturing cost per unit area to increase less (say, 30%) to mitigate the short-term risk. At the same time, the IC knowledge base is maturing, and the focus of ICM is changing toward foundry business and developing new product families based on the existing design and process portfolio in addition to exploring new technologies. All this makes ME DfM an attractive reference for improved profitability of IC DfM.

The intention of this book is to correlate the ME and IC approaches to DfM, understand their convergence path, and identify the ways to profit from this integrated knowledge pool, especially that IC DfM is growing in importance for the often razor-thin profit margins and may also become the key methodology to help IC business in the future. While IC DfM technical knowledge is becoming more commonplace, cost cutting can help with its expansion.

The basic engineering challenge is to build a product with a specific function using the shortest and lowest possible bill of the lightest and least expensive materials (BOM). ICM is not unique; on the contrary, it creatively extended this rule down to the nanometric dimensions, reducing the weight of the products with target engineering principles. But, a simple BOM may mean compromising reliability for weight. Therefore, trade-off is the name of the game in DfM. Dealing with trade-offs would also mean occasionally making a wrong judgment call. DfM should ensure the IC design makes no wrong turns when engineering the product.

DfM can be considered in two basic aspects:

Conceptual/business: How to maximize profits within a predefined technology of interest

Technical: How to define solutions to the existing manufacturing problems while increasing the value of the part

For ME assembly DfM, the following definitions have been proposed [5,6]:

Design for manufacturability is the process of proactively designing products to (1) optimize all the manufacturing functions: fabrication, assembly, test, procurement, shipping, delivery, service, and repair and (2) ensure the best cost, quality, reliability, regulatory compliance, safety, time to market, and customer satisfaction. It is evident that such a definition has to entail many more issues than just layout optimization by EDA. Accordingly, an important DfM practice is *concurrent engineering*, that is, the concurrent development of products and of their manufacturing processes. If existing processes are to be utilized, then the product must be designed for these processes. If new processes are to be introduced, then the product and the process must be developed concurrently. At the current stage, IC DfM is about both investing in the new processes and cost reduction and new applications of the existing ones.

Design for manufacturability and concurrent engineering for ME/assembly are proven design methodologies that work for a company of any size. Early consideration of manufacturing issues shortens product development time, minimizes cost, and ensures a smooth transition into production for quick time to market. These techniques can help commercialize the prototypes and their research. ICM still has a lot to learn from ME, which motivated this book.

1.2 Why Should One Read This Book?

Why should one read this book? After all, there are multiple IC DfM reviews, discussing how did this concept first emerge, and how does DFM now support design scalability according to the "more than Moore" paradigm [7]. Judging by recent publications, DfM trends and directions have been road mapped well, at least those in the key IC domain, microlithography. New pattern definition techniques are being developed based on heavy investment from EUVL, to immersion-based DPT (Double Patterning Techniques), to the more practical, rediscovered e-beam lithography [8].

But, this book is about something different from deep subwavelength layout and its consequences. It can claim to be the first superintegrated approach to DfM: The solutions relating to the different aspects of IC functionality are logically connected. It does not stay away from the pattern transfer subjects but treats them as one of many in the IC DfM domain. For the customer, a DfM subject more important than pattern transfer is reliability, which so far has not been addressed in DfM books. This is discussed in Chapter 2. While it is still true that IC DfM is largely about layout, one should first start from the way the layout is defined, as the primitive parametric cells or manually optimized memory arrays. Then, layout DfM needs to address how these cells are connected at the block and die levels to ensure low variability. This is covered in Chapter 3.

When reading IC DfM publications, one may be impressed that, were it not for EDA algorithms, IC technology as it stands now would not exist. This is not necessarily true, as shown in this book. To their own problems, manufacturing has a good number of internal responses, the most common of which is mask splitting. Only recently has this technique been widely used, in the form of DPT [9], to pattern ultrasmall geometries, for which EDA transformation of light and mask is helpless. In reality, DfM and MfD (manufacturability for design) compete with each other, based on trade-offs, for best return on investment (RoI). This is discussed in Chapter 4.

Many authors believe that the success of DfM should be measured [10]. Yet, a universal metric methodology, which would support and extend the commercial success of DfM, is not well developed and accepted. Chapter 5 offers for the first time a hierarchized system of DfM metrics from the quality of layout primitives, to hot spot forecasting at block level, to global ones such as yield and mask ratio, and their correlations. As the ICM techniques develop, the structure of this book should remain intact; only the underlying technical details would need to be updated.

1.3 The 2D Paradigm, Rules, and Optimization

It is not by accident that IC DfM has often been considered synonymous with layout for manufacturability, with quality driven by the ever-more-complex rule decks and systems. As long as ICs are designed and built in planar technology, the two-dimensional (2D) layout features define device performance and reliability when translated into a physical pattern on wafers. Recent attempts to introduce the third dimension into ICM based on die stacking and through-silicon vias (TSVs) are still based on 2D layout paradigms [11]. The resulting 3D integration still exploits a layered structure in which multiple steps eventually result in a 3D architecture. A fully 3D IC is hard to imagine at this point. Unlike other engineering products, such as automobiles, appliances, homes, and so on, the miniature IC does not have access from the third dimension. The only way to get "inside" an IC is by defining a manufacturing step that creates a cavity. In contrast, other mechanical systems are either built around cavities (homes, cars) or have cavities related to their function or architecture (washing machines). The 2D-to-3D gap makes it unsurprising that ME/assembly and IC DfM approaches were different throughout the years.

But, IC DfM is not only about adding new layout rules to the existing rule deck. It is a known disadvantage of design flow that if a design rule has not been proposed during the original process development and a failure mode due to the lack of such rule has been identified later, it is hard to insert such a rule into the deck as the design database (intellectual property, IP) may be invalidated. But, this methodology issue is not related to the criticality of the new rule versus the original ones. The occurrence of the failure mode due to the lack of a rule may be lower but the consequences are just as deadly.

At the same time, it is convenient and efficient to reduce DfM principles to three categories of rules: design guidelines (best practices), recommendations ("recommended rules"), and mandatory restrictions ("design rules"). As technology is maturing, the migration from guidelines to recommended rules to mandatory rules is concurrent with a geometric rate of increase in design cost (Table 1.1). New technologies first require mandatory rules; the negotiable rules and design best practices are helping with prospective improvements and cost reduction. For the subsequent technology nodes, recommended DfM rules and best practices often migrate into mandatory rules, resulting in increased design restrictions, which translates into higher complexity and cost, also due to the significantly increased total number of rules per IC mask layer from 3–5 at the outset to 10–15 at full maturity.

TABLE 1.1

Migration of Design Rules and Best Practices from Recommended to Negotiable to Required for the Subsequent Generations of Processes and Product Lines

	Scaling	Impact If Not Implemented	Process Generation	Category
Migration	Width Spacing Enclosure	Zero yield	−1	Required
	Directionality Well proximity Pattern density	Functionality corners	0	Negotiable
	Via doubling OPC reduction Symmetry	Non-competitive manufacturing	+1	Recommended

Although in contrast to IC DfM, ME DfM is often not based on a set of checkable rules, there can be a commonality between ME and IC DfM to drive cost reduction (Figure 1.1) (e.g., the number of defects). The x-axis in Figure 1.1 corresponds to a technical parameter that drives down the product cost. Device scaling was originally introduced to reduce the impact of particle contamination; smaller devices resulted in smaller product dice and higher yield at a fixed number of random point defects per wafer. However, improvements in particle yield were offset by the degradation in manufacturing yield, predominantly due to the patterning problems of small structures. Further reduction of critical dimensions (CD) resulted in a rapidly increasing manufacturing cost. To keep low CDs within the optimal process window, tighter process control was required, and pattern enhancements

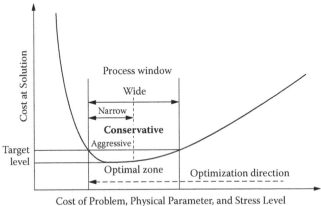

FIGURE 1.1
Universal process window optimization curve for lowest cost/maximum profit.

had to be monitored at a significant metrology effort. Therefore, aggressive DfM should be sponsored by high profit margins for the product.

Aggressive layout enhancement in the form of OPC (A-OPC) is typically used for the most dense regions containing memory cells, while conservative (sometimes called "minimum," M-OPC) or moderate (also known as "just enough," E-OPC) OPC is preferred for random logic. Relative areas of these blocks in an IC have a different impact on the die area. If C% of the die is covered by memory cells and $100 - C\%$ is covered by the random logic, one can assume that A-OPC enables area savings of C% times the aggressiveness factor $(100 - x)\%$, at the cost of

- exposure optimization effort
- physical verification runtime and complexity (database size)
- mask making (fracturing, write time, tools, inspection, etc.)
- delay to market due to these factors
- cost of manufacturing equipment, assuming no yield impact of the A-OPC

As a trade-off, A-OPC offers

- reduced die size (i.e., higher GDPW (gross die per water))
- increased device density for a wider range of applications
- improved functionality (e.g., speed)

DfM helps achieve positive RoI based on the cost-versus-time trend lines. The data points may be scattered, but if DfM optimization brings negligible benefits, the efforts are going in the wrong direction.

1.4 Three DfM Product Questions

DfM efforts are focused on both new products and expansions of the manufacturing base (e.g., foundry transfers). At design time, three DfM questions need to be answered before launching volume production:

1. Would the product work if it is built (functionality question)?
2. Can we build the product if we know it would work (prototyping question)?
3. Can we build the product at a profit (manufacturability question)?

While question 3 is most directly correlated to DfM, it may be asked too late if product viability is not first ensured. Because the cost of manufacturing grows fast with every subsequent step in actual product development, one should answer question 3 first, based on a paper study leading to a conceptual model. For example, in CMOS (complementary metal oxide semiconductor) technology, a literature search would readily show that while circuits with 1-nm-long transistors can be theoretically thought about, they have little chance of functionality, even if someone actually manages to build them. At the same time, paper studies are deficient, and simulations may fail; therefore, model margins for design should ensure that the concept would not become an implementation problem. The existing technology base may not suffice to provide product functionality across unexpected performance corners.

Figure 1.2 shows an example model of a product that encountered a DfM reliability problem, operating at a typical, high-speed/low-temperature performance corner. The design problem was due to the lack of behavioral models of maneuverability versus forward visibility at speed (related to a small rudder); suboptimal choice of primitive components (rivets) and material (low-grade cast iron); and lack of stress relief rules to prevent it from breaking into parts at joints with sharp corners, which, combined with insufficient redundancy of deck space (lifeboats) and insufficient wireless links to the redundancy available in the neighboring units to keep the customers in survivable conditions, resulted in failure analysis (FA) taking over 90 years at a multimillion dollar combined cost, offset by the revenue collected from public releases of FA reports in the form of high-profile motion pictures. Note that all DfM issues contributing to this quality incident were either readily known or easily predictable at the time of product release and were promptly corrected in the next revisions of the design.

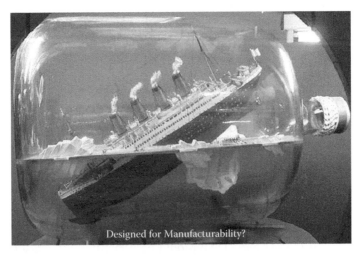

FIGURE 1.2
A high-complexity product misdesigned for reliability.

FIGURE 1.3
A product with lower complexity and functionality, with a better reliability image within its operating application window.

Startup companies are created in the belief that they can answer question 2. Ingenuous craftsmanship and process flow can create a working prototype even in a technologically challenging environment. Figure 1.3 shows a model of a simpler version of the product from Figure 1.2 with similar functionality. This product required low engineering skills and investment levels at the time of construction, but due to the more advanced DfM, it did not have as poor a reliability image as its great predecessor, if within its narrow operating parametric envelope, i.e., product specifications.

But, many of the startup companies go under if the answer to question 3 is "No." Among the reasons are not only poor yield or inadequate product appeal but also the limited market size, which cannot be resolved with engineering skills. All these issues call for a comprehensive definition of DfM goals. Figure 1.4 illustrates the key DfM challenge with volume production of the product shown in Figure 1.3. While the yield can be very high, the key issue is that the market may not absorb too many units. Therefore, DfM efforts would not translate into money. In contrast to many ME products,

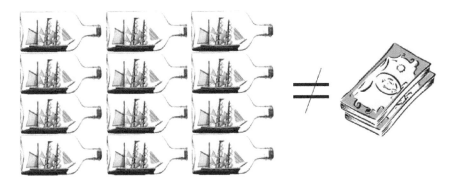

FIGURE 1.4
Can multiple products of simpler functionality gain enough market acceptance for a satisfactory profit margin?

DfM for IC helps satisfy the most pressing need of humankind: the need for control and information, which by all signs is in the expansion phase. Therefore, there would be IC products and processes for which the three DfM questions can be satisfactorily answered. This book should help select and manufacture them.

1.5 Smart Goals

The next step is to define SMART DfM goals. SMART is an abbreviation for significant, measurable, attainable, relevant, and timely.

1.5.1 Significant

DfM is an art of compromise; that is, for all the enhancements it promises, it carries a cost. At the beginning of the product development cycle, DfM can be a drag on the schedule. While the upside is that, once implemented, there would be no technical delays in volume production, DfM setup is an overhead, undesired for a product that has not yet shown its market potential. Therefore, management may prefer first to develop the prototype on a shoestring budget, followed by later retrofitting its design into DfM principles. To avoid this scenario, DfM should demonstrate its advantage as early in the design cycle as possible, such that all the key DfM improvements are readily captured in the prototype. One caveat is that DfM initial rules and guidelines would to a large extent be guessed or copied from the ones developed for the previous product generations. For example, one should expect that all existing IC layout manufacturability rules for the 28-nm process node would be first adopted based on the 45-nm legacy rules, and that this approach would not suffice. When the 28-nm node is in production, a concurrent DfM effort would be launched with foolproofing (e.g., Poka-Yoke [10]) focused on the existing, not on the previous, technology generation.

1.5.2 Measurable

No progress can be proven without proper metrics. For DfM, there are several of them; usually, all of them need to be employed to justify that DfM-related improvement is indeed "significant."

1.5.2.1 Manufacturing Yield

The favorite manufacturing metric is the yield Y. Its time dependence $Y(t)$ can be a measure of DfM success. However, given sufficient time or resources, approximately a 100% yield may be achieved, but the high cost (or poor

RoI) of yield optimization may take the company out of business. A subcategory of yield, called the first-pass yield (*FPY*) [11], may be a more relevant measure of DfM, but a high *FPY* can result from the extra engineering effort during processing of the prototyping material. Therefore, for IC DfM, supplementary metrics are required, such as the following:

Mask ratio (MR). This metric is defined as the total number of masks taped out to adjust product performance to the minimum mask count required for product functionality, and is well understood by the design community, reflecting product evolution in time. One can correlate the total design cost to the mask ratio based on the average redesign time and resources per mask layer and on the cost of material in line. The DfM mask ratio would be a fraction of the total mask ratio required to resolve manufacturing issues reflecting the overall ability of the fab to meet design needs.

DfM yield. DfM manufacturing yield and mask ratio are correlated by the DfM yield. In general, the higher the DfM yield, the lower the DfM impact on the manufacturing yield (Chapter 5).

RoIT (time to return on investment). A DfM solution that greatly enhances product quality may require a lot of time to implement and achieve benefits, and could miss the market window. Keeping track of DfM timeliness is critical.

1.5.3 Attainable

It would be unreasonable to expect that DfM could replace capital investment (e.g., new processes and equipment). No matter how advanced the DfM is, one cannot use the tool set of the past to manufacture products of the future. For example, manufacturing 65-nm products using 248-nm steppers may be a technical possibility, but only in a very limited range of applications. Layout restrictions may help produce memory matrices with 65-nm straight lines on a pitch of 180 nm, and if a product line develops a successful application for such a design, it could become a commercial success. However, under typical conditions (e.g., for random geometries at a pitch of 130 nm), no practical DfM would replace the reduction of exposure wavelength down to at least 193 nm, which requires equipment upgrades.

One should also not confuse DfM with new, more advanced engineering concepts. For example, new lithography techniques, nanoimprint lithography (NIL), and EUVL are not DfM but MfD, and in that capacity they may substitute DfM for pattern transfer from design to wafer.

1.5.4 Relevant

DfM techniques have been developed throughout generations of ICM processes based on different technical solutions, from MOSFET (metal oxide semiconductor field effect transistor) gates made of aluminum metal, to the self-aligned ones made of poly, and of refractory metals (e.g., tungsten, W), LOCOS (local oxidation of silicon), and STI (shallow-trench isolation), and the like requiring complex models. DfM tools and techniques should pertain to the actual manufacturability problems for the current technology node. Every technology has not only its specific challenges requiring new DfM solutions, but also secondary issues inherited from the previous technologies, for which solutions readily exist. For the IC design, the most complex mask layers (active, poly, contact, and metal 1) require leading-edge pattern fidelity tools. Other mask layers (e.g., defining large-area implants or higher-level metals) may use previous-generation DfM. Using model-based (MB) DfM when rule-based (RB) DfM would suffice, adds cycle time without adding value. In addition, MB corrections are more difficult to verify, compared to the RB ones, especially without a simulation engine and advanced metrology. In the event of re-tape outs, model resolution or computation tool algorithms may introduce random differences, the consequences of which take time and effort to quantify. For example, upgrading the fill pattern from RB to MB after the original masks are already built, makes it impossible to meaningfully run mask comparison without detuning the inspection to ignore fill geometries, and a real database defect can be unnoticed. Therefore, it is important to limit DfM development to the relevant minimum.

1.5.5 Timely

The timeliness criterion is the most arbitrary of all the SMART principles. One can illustrate this by the story of two men running away from a tiger. One of them hurts the other one's leg. The hurt man exclaims: "How is this going to help you? You cannot outrun the tiger anyway!" The other one replies: "Indeed, but now I can outrun you!"

Similarly, timely implementation of DfM should ensure that product introduction would not be delayed, not necessarily in the absolute timescale, but compared to what the market is expecting and what the competition is doing. Investing money in DfM tools may be difficult; investing engineering time "borrowed" from the future is easier. In fact, there may be nothing wrong with allocating more time for product development, but it is just as tempting to minimize DfM efforts trying to outrun the competition. However, this tactic can backfire at volume production, for which the competition, coming in at a later stage, can deliver in quantities and perhaps a more mature product. Of course, it is best to achieve both speed to market and volume production to fill the orders, so the trade-off has to be determined on a case-by-case basis.

1.5.5.1 *Verification Methodology*

Depending on the timing of technology development (early stage, midcycle, or late), one can propose three levels at which DfM can be applied (Figure 2.1).

Early in the technology development cycle, DfM becomes an enabling technique, without which the new product manufacturing cannot achieve the required functionality. As an example, scatter bars or phase-shifting techniques extended the life of optical lithography for successful designs at 65- and 45-nm technology nodes, using the same wavelength to print them as the one used for the previous technology generations.

With the maturing of technology, that is, at the **midcycle**, DfM shifts toward optimization of yield and device parameters. As an example, MB design (e.g., MB OPC) replaces RB design enhancements (e.g., RB OPC) to reduce the effort of verifying OPC metrology for arbitrary layout situations.

Finally, for **mature** technologies (i.e., at the late stage), DfM would help maximize the profit based on high product volume and small but consistent yield improvements. DfM priorities and expense commitments would migrate with technology maturity based on the concept developed for the ME/assembly DfM.

1.6 DfM: Is/Is Not

The scope of IC DfM methodology ranges from product architecture to fab yield to reliability; that is, it touches most aspects of product flow. This book focuses on high-level problem solving aligned with evolution of the IC industry. The intention is to provide a line of thought to help systematize the detailed technical information available from many sources. While recapturing this information in one publication would provide a book with a short expiration date, the history and methodology of DfM helps systematize the correlations between the technical and business solutions, applicable also in future situations.

Due to the breadth of DfM, it often escapes crisp definition [12]. Although there are many related disciplines, such as DfT (design for test), DfY (design for yield), DfR (design for reliability), LfD (litho-friendly design), and more, their correlation with DfM is not always clear. The following is a short list of is/is not issues typically falling under the generic DfM category (Table 1.2):

Is: Component Definition. The IC components, such as active or passive devices and design blocks, as well as their integration in the design flow, have an impact on the commercial success of the product. For system-on-chip (SoC) products with different functions enabled by one integrated circuit, one should select multiple components, such as memories, logic, high-voltage IOs (input/output), sensors, and actuators, designed according to conservative design rules, to reduce integration problems and improve reliability, model accuracy,

TABLE 1.2

DfM Is/Is Not

DfM	
Is	**Is Not**
… about component definition	… the result of device scaling
… about design for test	… about low power
… about system-level architecture	… the same as design for yield
… consistent with design for reliability	… but includes litho-friendly design
… dependent on design tools	… related to new devices

and physical verification of the layout. For SiP (system-in-package) products, one can use aggressive rules within the stand-alone components as long as the system architecture is consistent with their individual quality.

Is Not: Shrink Path. Moore's shrink path promotes DfM efforts, but these two concepts are independent of each other. For example, the photolithography roadmap enabled shrinking device dimensions from millimeters to micrometers to nanometers by making pattern reproduction rely on the many tool enhancements to mask writing, proximity correction, and so on, to drive the scaling. These efforts focused on minimum CD stability and reproducibility for a given technology generation, granted by the quality of the new equipment. Therefore, the shrink path is not a DfM methodology or vice versa, with one exception: By increasing the number of devices on a wafer, for a fixed intrinsic defect density, shrink path improves manufacturing yield. But, correlation between shrink path and DfM has limited impact as each technology generation suffers from its own yield loss mechanisms, and particle-limited yield at the current fab contamination level is not a major yield depressant.

Is: Design for Test. DfT is a subcategory of DfM because testing is a key enabler not only for parametric and functional yield but also for a product performing to customer expectations. Emerging device applications, new packaging options, and the related parasitic effects require reliable testing to understand every aspect of product operation. For example, if customer application of an RF (radio frequency) product creates data transmission disturbance due to poor PCB (printed circuit board) compensation, the IC maker should be able to rectify the problem within that application instead of falling back on product design changes and extensive testing at high cost.

Is Not: Low Power. This is device application independent of the manufacturability.

Is: System-Level Architecture. This is one of the aspects of correctness by construction (CBC). If the architecture is too complicated or does not meet market expectations, no manufacturing effort will be able to compensate for it.

Is Not: The Same as Design for Yield. Yield is a subcategory of manufacturability. One can have high yield without adequate manufacturability at a high price. The opposite is not true: One cannot claim good manufacturability without the required yield.

Is: Design for Reliability. According to the rule of 10, DfR is a prerequisite to DfM. If a product, otherwise manufacturable, is not reliable, it would have no value to the customer.

Is Not But Includes: Litho-Friendly Design. The terms *design* and *layout* are sometimes considered synonymous. But, design is a broad engineering category involving block architecture, models, and simulations, while layout is reduction of those design elements to mask level. LfD is a popular misnomer; it should more correctly be called litho-friendly layout (LfL) instead.

Is: Design Tools. One cannot have DfM without up-to-date design tools performing efficient logical synthesis, modeling, simulation (electrical, physical, optical, etc.), OPC implementation, and CBC layout. One cannot achieve manufacturability of next-generation devices with previous-generation design tools, even if Moore's shrink path is not being pursued.

Is Not: New Devices. New types of transistors, diodes, capacitors, etc., are introduced to broaden a product portfolio but not to resolve manufacturability challenges.

Is: Hot Spot Elimination. Design quality is verified by automated design rule checks (DRCs). Even for CBC design, it is hard to expect that DRCs would not be required. Depending on layout restrictions and model quality, DRCs can check the multiple aspects of IC manufacturability, such as reliability, functionality, lithographic yield, power domain definition, parametric yield, and so on. Early identification of hot spots related to all these disciplines would save months and cycles of learning, and there are several basic approaches to their elimination (Table 1.3; Figure 1.5). In this book, we discuss the best way to identify and eliminate design hot spots at different levels (Chapter 4).

TABLE 1.3

Approaches to Hot Spot Elimination

	Pros	Cons	Examples
Local repair	Quick/simple and may be sufficient	May cause domino effect May not address root cause	Manual layout changes
Reengineering	Consistent with CBC approach (e.g., using higher-quality IP (intellectual property) blocks) May apply to many products Addresses root cause	Time consuming If unproven, may create unknown problems	Adding a power domain Applying new fill pattern engine
Derating	May work on completed designs	Does not remove the actual defect Narrows the application (may lower the price point)	Reduce power supply Detune inspection tools

1.7 This Book: Is/Is Not

The preceding discussion explained the need for DfM synthesis and this book's approach to the subject matter. Accordingly, we will follow the product flow from architecture to the layout challenges and from DfM process definition, through the manufacturing, to the metrics. We will discuss several metrics to verify the efficiency of DfM approaches in the design cycle,

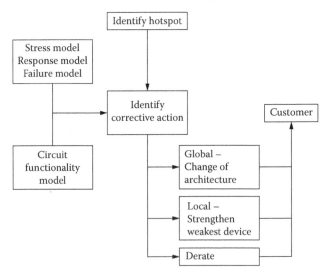

FIGURE 1.5
DfM optimization flow for hot spot removal.

which require the technical aspect to be subservient to the product performance and economy. In summary, this book

> *Is: This author's impression of an important discipline that propels the IC industry,* a methodological essay that should guide a patient reader through the layers of the problem.
>
> *Is Not: DfM Encyclopedia.* The included references highlight the argument but cannot be considered comprehensive, as their sheer number would overwhelm both the author and the reader.

This book is divided into six chapters. Following the introduction in this chapter, Chapter 2 guides us through the principles of DfM and shows how to align the universal DfM guidelines with IC engineering.

Chapter 3 discusses DfM at the device and product levels.

Chapter 4 reviews what DfM means for manufacturing and contrasts it with the alternative solution, MfD.

Chapter 5 presents the key DfM metrics of success.

Finally, in the summary in Chapter 6, the road map issues are discussed.

This book uses a number of technical terms. For ease in understanding, the most common terms appear in Table 1.4 and are defined again in separate chapters.

TABLE 1.4

List of Abbreviations and Key Terms

Term	Explanation
1D, 2D, 3D, 4D	1, 2, 3, 4 – Dimensional
(C)MOS	(Complementary) Metal-Oxide-Semiconductor
ABBA	Arrangement of active features BB between dummy features AA
ADC	Analog-Digital Converter
ASIC	Application Specific Integrated Circuit
ASP	Average Selling Price
ATE	Automated Test Equipment
ATPG	Automated Test Pattern Generation
BE	Back-End
BIST	Built-In Self Test
BoM	Bill of materials
CAD	Computer – Aided Design aka EDA (Electronic Design Automation)
CBC	Correct-By-Construction
CCM	Charged Cable Model
CDO	Carbon Doped Oxide
CDU	CD (Critical Dimension) Uniformity
CMP	Chemical-Mechanical Polishing (or Planarization)
CMPY	CMP Yield

(continued)

TABLE 1.4 (CONTINUED)

List of Abbreviations and Key Terms

Term	Explanation
COG	Chrome-on-Glass
CoL	Cycles of Learning
c_{pk}	Process capability index
CVD	Chemical Vapor Deposition
DfM	Design for Manufacturability – the subject matter of this book
DfR	Design for Reliability
DfT	Design for Test
DfY	Design for Yield
DoE	Design of Experiments
DPT	Double (or Dual) Patterning Techniques
DRC	Design Rule Check
DTI	Deep Trench Isolation
DUV	Deep Ultraviolet
EDA	Electronic Design Automation
EEPROM	Electrically Erasable/Programmable Read Only Memory
EFR	Early Failure Rate
EOL	End-of-Life or End-of Line
EPE	Edge Placement Error
EUV(L)	Extreme Ultraviolet (Lithography)
FE	Front-End
FET	Field Effect Transistor
Fill pattern	Wafer pattern not performing design function, required only for technology reasons. A.k.a. waffles (waffling), dummy fill
FINFET	MOSFET with the channel wrapped by a thin silicon film
FIT	Failure in Time
FMEA	Failure Mode and Effect Analysis
FN	Fowler-Nordheim (oxide tunneling) current
FSM	Finite State Machine
GDPW	Gross Die Per Wafer
GDS	Graphic Database System (data file format)
GIDL	Gate-Induced Drain Leakage
g_m	MOSFET Transconductance
HCD	Hot Carrier Damage
HCI	Hot Carrier Injection
HD	High Density
Hotspot	Contextual defect of layout, allowed by nominal design rules but limiting reliability, functionality, or yield
I_{dsat}	Drain current in saturation
ILT	Inverse Lithography Technology
IO, I/O	Input-Output

TABLE 1.4 (CONTINUED)

List of Abbreviations and Key Terms

Term	Explanation
IR, IR drop	Voltage drop due to current I flowing through resistor R
ITRS	International Technology Roadmap for Semiconductors
I-V	Current-Voltage
LDD	Lightly Doped Drain
LEE	Line End Extension
LER	Line Edge Roughness
LfD	Litho-friendly Design
LfL	Litho-friendly Layout
LFR	Late Failure Rate
LNA	Low Noise Amplifier
LOCOS	Local Oxidation of Silicon
LOD	Length of Diffusion
L_{poly}	Channel Length (poly gate width)
LVL	Layer vs. layer: layout comparison algorithm
LVS	Layout vs. Schematic
Mask	Lithographic stencil used to transfer design pattern to wafer. A.k.a. reticle (due to fine grid of lines / spaces), photomask
mb	Model-based
MDP	Mask Data Preparation
MEMS	Micro-Electro-Mechanical Systems
MfD	Manufacturability for Design: a methodology to meet design goals by investing in fab process or equipment
MR	Mask Ratio
MRAM	Magnetic Random Access Memory
MRC	Mask Rule Check
MTBF	Mean Time Between Failures
MTTF	Mean Time To Fail(ure)
NBTI	Negative Bias Temperature Instability
NGL	Next Generation Lithography
NIL	Nano-Imprint Lithography
NVRAM	Non-Volatile Random Access Memory
OPC	Optical Proximity Correction
O-TFT	Organic Thin Film Transistor
pCells	Design cells with parametric control of design features
PCT	Pressure Cooker Test
PL	Product line(s)
PnR	Place-and-Route
PR	Photoresist
PSG	Phosphoro-Silica Glass
PV	Physical Verifications
PVD	Physical Vapor Deposition

(continued)

TABLE 1.4 (CONTINUED)

List of Abbreviations and Key Terms

Term	Explanation
rb	Rule-based
RC	Resistance-Capacitance
RCP	Reliability Critical Path
RDL	Routing Distribution Layer
RET	Reticle Enhancement Techniques
RF	Radio Frequency
RIE	Reactive Ion Etching
RLC	Resistance-Inductance-Capacitance
RMS	Root Mean Square
RoI	Return on Investment
RTL	Register Transfer Level
S	MOSFET Subthreshold Swing (or slope)
SAR	Successive Approximation Register
SB	Scatter Bars
SILC	Stress-Induced Leakage Current
SiP	System in Package
SMO	Source-Mode Optimization
SoC	System on Chip
SPICE	Simulation Program with Integrated Circuit Emphasis (also: hspice)
SRAM	Static Random Access Memory
STI	Shallow Trench Isolation
TDDB	Time Dependent Dielectric Breakdown
Transistor NPN	Bipolar junction transistor with n-type emitter/collector and p-type base
Transistor PNP	Bipolar junction transistor with p-type emitter/collector and n-type base
VCO	Voltage Control Oscillator
VDD	Power supply (pin) in IC's
Verilog	Hardware Description Language
V_G	Gate Voltage
VHDL	Very High Speed Integrated Circuit Hardware Description Language
V_{ss}	Ground (pin) in IC's
V_T, V_{TH}	MOSFET Threshold Voltage
W/S/E	Width/Spacing/Enclosure - basic design/layout rules
WLBGA	Wafer Level Ball Grid Array
XOR	Exclusive OR

Throughout this book, abbreviations should define a technical term in an unambiguous way.

References

1. A dee, S. Top 10 facts gleaned at EUV Litho Workshop. http://spectrum.ieee.org/tech-talk/semiconductors/devices/top-10-facts-gleaned-at-euv-litho-workshop, 2009.
2. Moore, G. Cramming more components onto integrated circuits. *Electronics*, 38: 114–117, 1965.
3. Anderson, D. M. *Design for Manufacturability and Concurrent Engineering*. Cambria, CA: CIM Press, 2008.
4. Orshansky, M., Nassif, S. R., and Boning, D. *Design for Manufacturability and Statistical Design: A Constructive Approach*. Springer Series: Integrated Circuits and Systems 14. New York: Springer, 2008.
5. Bralla, J. G. *Design for Manufacturability Handbook*. New York: McGraw-Hill Handbooks, 1998.
6. Helander, M., and Nagamachi, M. *Design for Manufacturability*. New York: International Ergonomics Association, Taylor and Francis, 1992.
7. Tummala, R. R. Moore's law meets its match. *IEEE Spectrum*, 6, 2006, 44–49.
8. Wong, N. P., Mittal, A., Starr, G. W., Zach, F., Moroz, V., and Kahng, A. *Nano-CMOS Design for Manufacturability: Robust Circuit and Physical Design for Sub-65 nm Technology Nodes*. New York: Wiley-Interscience, 2008.
9. Wiaux, V., Storms, G., Cheng, S., and Maenhoudt, M., The potential of double patterning immersion lithography for the 32nm half pitch node, http://www.euroasiasemiconductor.com, 2007.
10. Aitken, R. DFM metrics for standard cells. *Seventh International Symposium on Quality Electronic Design*, March, 491–496, March 2006.
11. Reisinger, J. 3D design challenges—from WLB to PoP. IFIP/IEEE VLSI-SoC, 16th International Conference on Very Large Scale Integration, Rhodes Island, Greece, October 2008.
12. Chiang, C., and Kawa, J. *Design for Manufacturability and Yield for Nano-Scale CMOS/Integrated Circuits and Systems/*. New York: Springer, 2007.

2

Migrating Industrial DfM into IC Manufacturing

2.1 Introduction: Universal Design for Manufacturability Principles

Since the late 1800s, engineering industries (construction, ship- and rail-based transportation, electrical engineering, and many others) have developed a legacy of design for manufacturability (DfM) guidelines and methods, further referred to as mechanical engineering (ME)/assembly DfM. They became known as the rule of 10, the rule of 80/20, concurrent engineering, correct by construction (CBC), Pareto distributions, and so on [1]. Many of them were recognized by the semiconductor industry in more or less conscious ways. This chapter discusses why and how integrated circuit (IC) manufacturing should take better advantage of them, based on examples related to improving the quality of IC product design. Because DfM does not generate new product functionality, its main value is cost reduction, reflected in three phases (Table 2.1):

- Initial: minimizing the cost impact of product flow setup
- In production: reducing the cost of manufacturing
- In the field: avoiding the cost of reliability

The relative importance of each of these phases is different from the designer and the customer points of view. For the designer, the most important is the initial phase. The direct cost of product flow or project setup is usually low, but it has a significant impact on product viability. Cost expectation for advanced technologies is split in an 80/20 ratio between the cost commitment and the execution phases of product life (Figure 2.1).

Because a significant fraction of the product cost is readily committed in the first phase, taking advantage of the legacy ME/assembly for IC flow setup methodology should help drive cost reduction early in the process. The defining phase of the IC flow, concurrent with the architectural definition, also requires setting up the bill of materials (BoM) similarly, as it is

TABLE 2.1

Time-Dependent DfM Cost Categories and Rules

Cost Category	Origin	DfM Rule	Explanation
Setup (initial)	Design	80–20	"Do the right thing" 80% cost committed at 20% into product lifetime
Execution (in production)	Manufacturing	CBC	"Do things right" Correct by construction more cost effective than testing
In the field	Reliability	Rule of 10	"Failure is not an option" The later the defect found, the larger the financial impact

done for a mechanical product (selecting nuts and bolts). Choosing the right primitive devices (MOSFETs [metal oxide semiconductor field effect transistors], diodes, resistors, etc.) is critical for the subsequent logical synthesis, timing closure, and design verification (physical rule checks); therefore, these devices must be supported by robust models. Defining the placement

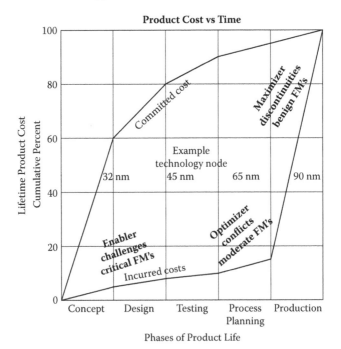

FIGURE 2.1
Phases of DfM depending on the phase of product life, corresponding to example technology nodes and aligned with failure mode (FM) severities. (Adapted from Anderson, D. M., *Design for Manufacturability and Concurrent Engineering.* Cambria, CA: CIM Press, 2008.)

design blocks of IP (intellectual property) blocks in the IC (integrated circuit) architecture can be compared to the subassembly of a mechanical product. Transferring the design to the IC manufacturing line would be an equivalent of the final mechanical assembly, with poor yield indicating mismatch between expectations embedded in the models of the components and their interactions, and the testing conditions. The reason why the entire IC manufacturing line can be compared to the final mechanical assembly phase is that there are no design changes permitted once the IC design is manufacturing ready for volume production. The watershed phase of IC product flow is pattern transfer from the conceptual world of design data to the physical world of masks (tape out) and then fab. Parametric data of the product cannot be adjusted incrementally after this step, and there are only a few opportunities to undo incorrect manufacturing operations.

Based on this comparison, DfM approaches for mechanical products and for the ICs can also be correlated. The methodology calling for high cost commitment early in the product lifetime makes it critical for both ME/assembly and IC DfM to ensure that the product architecture is correctly defined, that is, the product is "doing the right thing." Product definition is often driven by intangible factors, such as inspiration in hope of the anticipated market demand, or by superior quality and reliability of a product performing a legacy function. In the latter case, a conservative approach, such as relying on the improved perception, is often successful. Even if the new product would not provide any extended functionality compared to the existing one, but the legacy functionality is performed in a safer, more reliable and comfortable way, it would still enjoy success, similar to that enjoyed by steamships over tall ships, electronic calculators over an abacus, and systems on chip (SoCs) over mechanical controllers.

Standardization and concurrent engineering as well as other ME/assembly DfM legacy principles are helpful at the execution stages of physical design of the product flow (Table 2.2). Because manufacturing cost accumulates with time (Figure 2.1), DfM should be committed and introduced as early as possible for maximum advantage. It may be difficult to introduce DfM in the conceptual phase of the product flow because the new inventions have to be based on manufacturing experience, and DfM has to rely on mature tools and reliable processes. For designs proven on prototypes, built and adjusted by the engineering teams, DfM must subsequently not only bridge the gap between design, technology, and product performance but also cater to customer needs, to which the factory and its suppliers manufacture. For the customer, the most important aspect of DfM is product quality over its lifetime. Therefore, in the product setup the key goal of DfM is to avoid quality incidents. For a high volume of products with multiple applications, designs working 99% of the time are not good enough. All products should work as required by customer applications, throughout their specified lifetime, to address the first critical question of manufacturability: Would it work if we build it (i.e., Did we do things right?)? At the same time, even conservative DfM should aggressively help increase the profit margin driven by the manufacturing yield.

TABLE 2.2

Phases of DfM Implementation and Their Weighted Impact

Weight	Category	Features
Testing		
1	Parametric	Fast but not comprehensive
10	Functionality	Complex and expensive
100	Reliability	Slow, very limited sampling, impossible to test all
Electrical Design: Yield Impact		
1	Parametric	OPC, CMP, CAA, matching, via doubling
10	Functionality	Device models, IP
100	Reliability	HCI, ESD, latchup, EM
Physical Design: RTL to GDS		
1	Definition	Correct by construction
10	Execution	Design intent, tolerance bands
100	Verification	Hot spots

When translating a legacy DfM approach from ME/assembly to IC DfM, three questions need to be answered (Table 2.3):

1. What is the product made of? Depending on the product family and applications, DfM would deal with different materials and reliability issues.
2. How is it built? Microscopic and macroscopic design simulation and verification tools use different models and approaches.
3. What are the product operating conditions that have an impact on reliability expectations? For the ME products, it is the number of moving parts, while for the IC products, it is the operating voltage and signal frequency, which makes the biggest impact.

TABLE 2.3

Legacy DfM Principles

	Rule or Best Practice	Area of Importance
1	Rule of 10	Setting up product architecture Task prioritization
2	Correct by construction	Reduction of testing
3	Concurrent engineering	Shorter time to market
4	Doing it right the first time	Cost reduction Shorter time to market
5	Standardization	Shorter time to market Quality assurance Component and material selection for models and physical space

TABLE 2.4

Phases of Rule of 10 in IC Lifetime

Level	IC Product Cycle	ME Assembly Stage	IC Verification Method	Cost Impact
1	IC design	Component	Models	1
2	Design verification	Subassembly	Physical	10
3	Fab	Final assembly	Parametric verification	100
4	Test	Dealer/distribution	Functional verification	1,000
5	Customer	In the field	Reliability	10,000

The answers to these questions should ensure success in the marketplace controlled by the rule of 10 in the final DfM phase.

2.1.1 The Rule of 10

The key principle of ME/assembly DfM is known as the rule of 10 (Table 2.4). It says that the cost of design defects increases tenfold with each subsequent phase of product life, from inception to obsolescence (Figure 2.2). The rule of 10 emphasizes reliability as the most important aspect of manufacturability: A manufacturable product must not fail because the cost of failure could not

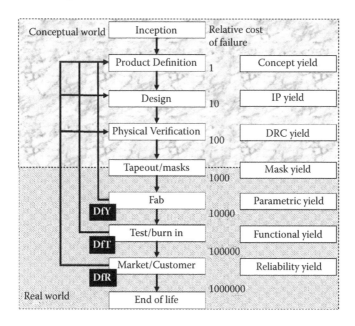

FIGURE 2.2
Product flow stages and different types of yield corresponding to the rule of 10 throughout product lifetime and the associated types of yield.

only be higher than the direct cost of development and manufacturing but also exceed the entire product revenue.

The rule of 10 helps hierarchize DfM implementation across key areas of interest: design setup, yield, testability, functionality, and reliability, the first ones having the highest visibility to the product line, but of the lowest importance to the customer, and vice versa. But, to satisfy the customer and ensure low product cost, one should work backward from the end of product life throughout its reliability and performance in the field, to its functional and parametric yield (PY), testability online, and finally design rule (DR) methodology, simulations, and design tools.

The cost increments corresponding to the subsequent IC development and manufacturing phases p_i, $i = 1, \ldots j_n$, each consisting of i_m substeps aggregate during the lifetime of the IC DfM product (Figure 2.3). To eliminate the parts that can incur the incremental cost of defects early in the product flow, the rule of 10 would stipulate introducing tests at significant checkpoints, defined by

$$\text{Cost}\left(\sum_{i=1}^{j_m} p_i\right) \leq 10 \tag{2.1}$$

Term	Explanation
p_i, $i = 1 \ldots n$	Manufacturing phase: setup, design, verification, tapeout, etc.
i_m	Manufacturing step at which testing is recommended
m	Number of steps into which one can subdivide the design/manufacturing process

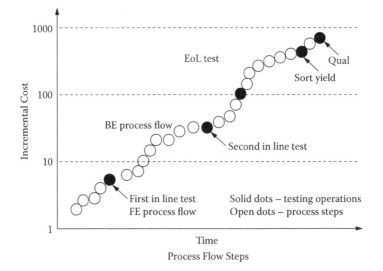

FIGURE 2.3
Incremental cost of testing steps within a hypothetical process flow (EoL = end-of-line BE = back-end; FE = front-end).

In other words, the incremental process steps over which the product cost increases by a factor of 10 with respect to the previous checkpoint could be considered candidates to become the next test checkpoint in the flow. This way, the risk is reduced to 10 times the amount invested in the preceding already verified steps. Models of the manufacturing line should help determine the step i_m for Formula 2.1. One should also set up test criteria at such 10× product cost milestones. Alternatively, because the product testing consumes resources and causes delays, one may chose to test products less often (e.g., at the 100× milestones). Between the testing steps, the process is considered CBC, with quality ensured by procedures, components, or tools, and reliability proven by independent studies. Reducing testing frequency introduces DfM risk, and reducing that risk has to be the key task of CBC engineering, as a key principle of ME/assembly DfM ported to IC DfM (Table 2.5).

The efficiency of CBC in concurrently eliminating both IC defects and testing operations depends on the quality of the models. To build a device that realizes a function demanded by the market and operates without interruption until the end of its specified life, one has to be able to predict all aspects of its operation. Ideally, with CBC-based component selection, a single test point at product setup time or any other testing step n in the product flow that would provide meaningful go/no go information should ensure 100%

TABLE 2.5

Pros and Cons of the CBC Approach for the Different ICM Disciplines

	Discipline	Advantage	Disadvantage
1	Modeling	No model interpolation/ extrapolation permitted for unsupported options	Performance may depend on layout context
2	Design	Improved accuracy of simulation using discrete models	Limited device options for optimization of parasitics
3	Layout	Reduced variability, simplified placement	Arbitrary, sub-optimal array efficiency
4	OPC	Improved environmental control	May need context-dependent optimization within pCells (parametric cells)
5	Quality	Reduced variability, simplified characterization	Context dependence
6	Manufacturing	Consistent process targets, simple metrology	May be compromised by pattern density distribution
7	Product line	Clear trade-off: area vs. quality	Default footprint may be too large
8	Mask shuttling	Improved defect detection	Systematic in errors may be unnoticed in (DTD [die-to-die] inspection)

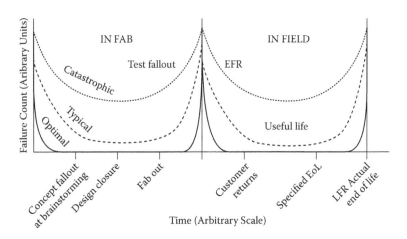

FIGURE 2.4
Combination of bathtub curves showing in-fab and in-field yield fallout (EFR = early failure rate; LFR = late failure rate).

yield at the outer end (EOL) of the bathtub curve (Figure 2.4). Fallouts at the following $n + 1$, $n + 2$, and so on, steps would be 10^{n+1}, 10^{n+2}, and so on, times more expensive and reduce the return on investment (RoI) of the single-point testing. On the other hand, DRCs and in-line testing would help verify the CBC approach, but the challenge is to minimize their impact on the cost of manufacturability.

Example 2.1

A team of designers would need 1 man-year to develop a new IC product. Without CBC, the team would subsequently require about four times more effort for design verification (i.e., 4 man-years). If the verification missed a defect, a problem may be found in the fab (e.g., within 2 months of wafer processing). Design debugging and testing would then delay the life cycle of the product (e.g., by 1 year [or by 10%] assuming a 10-year lifespan) and decrease the revenue accordingly. In the event of reliability issues at the customer site, the cost of the recall of the shipped material would combine with the loss of credibility. The resulting loss could match or exceed the total product revenue over its lifetime (Table 2.6). The lifetime of the IC products is often assumed to be 10 years, which is 10 times longer compared to the standard manufacturer's warranty. In other words, the IC products are often not built to last over 10 years, although many of them may function for a much longer time, on the one hand reducing the consumer's motivation to replace them, but sending a good quality message on the other.

2.1.2 Concurrent Engineering and Standardization: Design, Process, Product

The next principle of legacy ME/assembly DfM is concurrent engineering [1], which should help find the highest RoI point between standardizing and

TABLE 2.6

Illustration of Rule of 10 in IC Product Flow

Level	IC Product Cycle	Examples of Original Cost Components of Design Production or Delivery		Cost of Correction	
		Effort/Cost	Value	Effort/Cost	Value
1	IC design	1 man-year	200	1 man-day	1
2	Design verification	4 man-years	800	2 man-weeks	10
3	Fab	1 lot/week for 5 years 5,000 wfr at $500/wfr	2,500	2 months of manufacturing	100
4	Test	$100/wfr	500	10% of revenue	1,000
5	Customer	Overhead $200/wfr	1,000	Recall and image loss	10,000
	Total Cost		5,000		
Total revenue	500 wfr at $2,000/wfr		10,000		

Note: Values are in arbitrary units. Here, 1 man-day was set as 1 unit. Depending on when the defect is noticed, the impact propagates 10-fold for every next step.

customizing design and manufacturing flows, which is often more important than unqualified innovation. To gain market share, product lines add a new part or process to the existing portfolio, with as many elements of the old product flow as possible kept unchanged. The key issue is how to efficiently spin off a new value-added process with minimal investment and entropy. Due to the IC price erosion related to engineering advances, semiconductor companies have to generate new products to sustain their stream of revenue. Following Moore's law [2], device density increases by about two times every 2–3 years, so one should expect a price reduction by a similar factor forced by the competitiveness. Therefore, nonmanufacturing divisions of IC companies (research, design, and product lines) must provide the make-up revenue and, preferably, growth. This can be achieved by customizing standard processes toward new applications or by standardizing custom processes or products, implementation which have no history of success but have potential to provide competitive advantages in the emerging markets. One can propose a rule of thumb, by which existing standard processes should be 50% customized or new processes should be 50% standardized, to provide a maximum ratio of benefit to cost simply by staying far enough from the "all-standard" and "all-custom" extreme options. In manufacturing, it is beneficial to use a common process and fab site for multiple technologies by standardizing mask data preparation (MDP) algorithms and process recipes. On the other hand, incremental revenue can often be won by diversifying a product portfolio through custom products. One example of such an approach is memory products with different array densities.

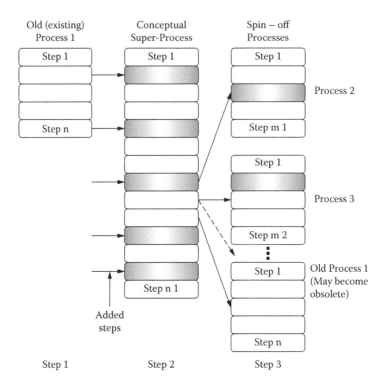

FIGURE 2.5
Concurrent engineering methodology of spinning off multiple product flows from a common baseline.

In mature industries (i.e., the ones with limited impact of innovation), core product parameters are readily defined, and design efforts are focused on improving product economy: the cost of ownership or manufacturing. The profit margins are often razor thin and based more on market trends than on product virtues. Changes enabling higher revenue are introduced at as few as possible critical steps to ensure risk reduction and cost control. The reference baseline, which we call process 1, after adjustments would become a derivative, which we call process 2, and starts living its own life (Figure 2.5).

The desire to keep both old process 1 and new process 2 as common as possible is reduced over time due to the potential new product opportunities for process 2 and end-of-lifing of process 1. Sometimes, process 2 is transferred to a different manufacturing site with the desire to keep cross compatibility ("copy exactly" as many steps as possible) with process 1. But, the differences in the equipment sets may deny many benefits of the compatibility.

One should note that standard flows for different products run in one fab, or for the same product run in different fabs (manufacturing lines), would eventually diverge to maximize yield. At the same time, their maintenance may become a significant fraction (e.g., 50%) of the cost of the original flow, which would indicate the need for retro-standardizing. However, the copy exactly methodology may work only in fewer than 50% of cases, subsequently requiring more resources for process alignment. Lowest cost (cost of entropy) is achieved by a flexible customize/ standardize approach based on documentation comprehensive enough to understand the differences among the flows and the consequences of these differences. Cloning processes and their documentation should start from creating a conceptual superprocess, comprehending all the steps and their variations in one core flow and in its specification (Figure 2.5). This core flow would not be executable but would form the base from which the individual, dedicated processes are to be branched off. That way, each step would be represented only once in the core flow and copied into its own branch, from which it is executed. The alternative approach, linear additions of steps directly into the custom flows, makes it difficult to run cross comparisons of processes and documents and understand the consequences of all the variants.

2.2 Design-for-Reliability: Reliability First

2.2.1 Definitions

The consequential cost of poor reliability (i.e., of customer disappointment due to product failure in the field) puts design for reliability (DfR) methodologies at the top of DfM priorities. High cost of testing requires that DfR would heavily rely on CBC and model development.

Successful DfR requires first a clear definition of reliability, aligned with market expectations. Reliability is driven by statistics and probability and is monitored throughout the entire life cycle of a system, including development, test, production, and operation, as follows:

$$R(t) = P(T > t) = \int_{t}^{\infty} f(x)\,dx \qquad (2.2)$$

Term	Explanation
R	Reliability (time dependent)
$f(x)$	Failure probability density function
T, t	Time (assumed to start from $t = 0$ at product definition)

- First, reliability, measured as probability of success at a statistical confidence level, for IC products is often set as 10 FIT (failures in [specified] time, usually 10 years). It is therefore acknowledged that the devices might fail.

- Second, reliability is operation of a part (i.e., a system or its critical subsystems) without failure with respect to its intended or specified function. And, even if no individual part of the system fails but the system does not perform as intended, it still has an impact on part reliability. Finding and fixing individual reliability failing points (hot spots) may not be sufficient to call a design reliable. System-level simulations across performance corners are necessary.

- Third, reliability applies to specified terms, for which a system has a predetermined chance to operate without failure within the assumed time, mileage, or cycles of use.

- Fourth, reliability is restricted to operation under stated, not unlimited, conditions, which one should be able to simulate. The product should also have the ability to "fail well," without catastrophic consequences.

Reliability level and budget may depend on the consequences of failure. A common reliability parameter, the mean time between failures (MTBF), or the number of failures during a given period (also known as the failure rate), is not useful for a subsumed parameter of ICs as they are usually not repairable. In some cases, IC reliability may be specified as the probability of mission success, for a single-shot device or system, maintained in sleep mode and only operating once in its lifetime, such as ICs in automobile airbags, thermal batteries, or missiles. But, in most common ICs, the legacy reliability parameters defining DfR do not apply and are replaced with MTTF (mean time to failure). The early failure rate (EFR) determines the failure distribution with a decreasing failure rate over the first part of the bathtub curve, under moderate stress for a limited period of time (a censored test), to identify the most marginal hot spots. The EFR is often pessimistically modeled with a one-parameter exponential distribution for a small sample size. For the stress agents, stress time, or the sample size so low that not a single failure occurs, zero-defect experiments determine an upper limit of the EFR. While it may look good for the customer that there are no failures, the confidence level is low. In IC design, it is similar to "violator cell" verification test cases (i.e., a controlled environment), which, while important to setup reliability flow, is insufficient for volume production. To determine the intrinsic failure distribution (e.g., due to material properties), higher stresses are necessary to induce a failure in a reasonable period of time. Several degrees of stress have to be applied to determine an acceleration model as testing under nominal conditions would take too long (e.g., 10 years). The empirical failure distribution is often parameterized with a Weibull or a log-normal model.

Because reliability is a probability, even highly reliable systems have some chance of failure. The role of DfR is to reduce the failure margin to insignificant limits and ensure noncatastrophic failures by randomizing failure modes.

Testing reliability requirements has many limitations. A single test is insufficient to generate statistically significant data, but multiple or long-duration tests are expensive. The goal of reliability modeling is to design

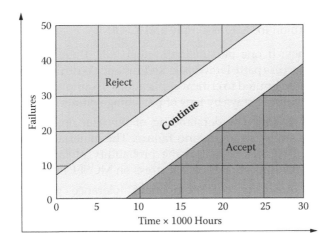

FIGURE 2.6
Time and failure-count-dependent product disposition. A reliability sequential test plan. (Based on MIL-STD-785, IEEE 1332. From the U.S. Department of Defense 1980. Reliability Program for Systems and Equipment Development and Production.)

a test program, which provides enough evidence that the system meets its requirements. Parameters are listed along with their corresponding confidence levels (e.g., MTTF of 1,000 hours at 90% confidence). From this specification, a test can be designed with explicit criteria about the number of hours, number of failures, and cost and risk to both the customer and the producer (Figure 2.6). Reliability testing may be performed with extreme temperature, humidity, shock, vibration, and heat conditions [3].

DfR models should identify devices (e.g., based on their layout geometries) that carry potential risks to reliabilities expressed by the desired MTTF, FIT, and confidence values calculated manually or with simple point tools. The accelerated life testing in the laboratory should then induce failures in such devices at a much faster rate than in nominal operating conditions by providing a harsher, but still representative, environment. The product would fail in the lab just as it would have failed in the field, but in a much shorter time. The models use several types of equations (e.g., Arrhenius, Eyring, inverse power law, temperature-humidity) to prove their conformance to the bathtub reliability curve, linked to Weibull analysis and statistical interference.

Verification of reliability includes:

- Built-in test (BIT) (optimal from a DfM standpoint) with accelerated testing
- Fault tree: identify the device that causes the circuit function to fail, linked to the avoidance of single-point failure (hot spots)
- Sneak circuit analysis

The goal of DfR is to ensure product conformance to its specification, using techniques common for ME/assembly and IC DfM:

- Redundancy: If one part of the system fails, there is an alternate (backup) success path. Drawback: Redundancy is difficult (expensive) and therefore limited to critical applications. Example: Improving the yield of memory arrays by tens of percentage points (Table 2.7).

- Modeling: Detailed understanding of physical processes with an impact on stress, strength, and failure. The material or component can be redesigned to reduce the probability of failure. Example: Reducing the impact of electrical stress on MOSFET parameters.

- Derating: Ensuring that component tolerance withstands the expected stress by using a heavier-gauge component, which exceeds the normal specification. Example: Increasing MOSFET channel length to reduce hot carrier effects.

Reliability testing for complex systems may be performed at many levels, such as device, circuit board, unit, assembly, subsystem, and system. They all require built-in circuits or test chips to improve confidence at the device level, which needs to be communicated to customers to select the best option, which then becomes a part of the product specification.

A key aspect of reliability testing is to define *failure*. It may not be clear whether a failure is really the fault of the system or a result of variations in test conditions, operator skills, ambience (weather), and unexpected situations. The rule of 10 and reliability modeling should help eliminate the most consequential failures starting from the most expensive ones, when the product is already in the customers' hands, back to the least-expensive ones, for the product in the definition stage.

TABLE 2.7

Example Estimates of Cost of Redundancy Indicating Complex DfM Tradeoffs

Cost Factor (% of Total Design)	SRAM Redundancy	Via Doubling (%)
Die footprint increase	2%	0
Design cost increase	5%	1
Layout cost increase	2%	0
Repair equipment	$1 million	0
Repair process	$5/wfr	0
Retest after repair	$5/wfr	0
Negative yield effect	0	2
Controllability	Yes	As programmed
Total yield improvement	25%	7
Total cost	$1 million/tool	2
	5% design	
	2% + $10/wafer	1

Example 2.2

The highest cost of failure in the field (e.g., 30 man-years or 360 people working for 1 month) should be reduced to the cost of the design correction (e.g., of 1 man-day) by making the right component choice (Table 2.6).

2.2.2 Material-Dependent IC Reliability

IC reliability primarily depends on the mechanical (physical) and electrical properties of the material components of the circuits. The mechanical properties are modified by the physical stress, while the electrical ones are modified by the voltage and current stimuli applied to devices and interconnects (Table 2.8).

Mechanically, the ICs usually have no functions other than self-support, and a stable form factor is the guarantor of their proper operation. This may change in the future, as the lack of movable parts is only true for the ICs not being used e.g., as motion sensors, with their packages strong enough to withstand the environmental pressure, light, humidity, and temperature within the operating range. For ICs printed on a sheet of plastic, the reliability of active and passive devices would more directly depend on the external form factor. At this point, reliability of an IC is a function of layout geometries inside the die and can be modeled first at the individual device level. A DRC can find devices in the die schematic or layout that compromise system reliability. The DfR criteria are addressed to geometries being able to withstand the expected electrical and mechanical stress levels required by the product function and to CBC building blocks, which reduce the risk even before the DRC checks (Table 2.9). Many concerns about die performance related to mechanical

TABLE 2.8

Material-Dependent Reliability Issues

Stress Type/ Failure Mode	Metals	Dielectrics	Semiconductors
Physical/ mechanical	Cracking propagating from top surface Delamination	Cracking (e.g., layers covering active under pad)	Mobility variation Threshold shifts
Chemical	Corrosion due to humidity ambient	Corrosion due to penetration of residual etch agents (e.g., HF)	Contact spiking, (Al/Si reaction)
Electrical	Electromigration, IR drop on power grids (DC [direct current] or RF)	Electrical overstress (HCI, NBTI, ...) induced charging or breakdown (ESD, TDDB)	Latchup (SEL/SER) power noise, ESD
Enhancement factors	Humidity, temperature, pressure		Radiation, temperature

TABLE 2.9

Failure Prevention at Layout Using CBC

Domain of Layout Best Practice	CBC Approach
Connectivity	Full connectivity for all hierarchy levels
	No extraction errors
Power grid quality	2D grid in all blocks with subsequent metals perpendicular to each other
	Redundant lines in top 4 metals
	Pad connected to diffusion away from injection-sensitive areas (diffusion, high-Z resistors)
Sensitive regions	Analog, high-Z, sensitive regions must have shielding
	No parallel lines longer than 5 μm or 2% of length
Matching	Use dummy–ABBA*–dummy FETs, LOD (length of diffusion), orientation, gate extensions, RLC
	Same orientation of FET source-drain
	Dummies tied to the same node
DfT	Use LVSable probe points to top metal on critical nodes
Wells	Reduce areas for low impact on die footprint
Line W	Line W > 2× min CD for wide devices, 4× for ultrawide,
Contact count	use 8 contacts or vias
Latchup	Substrate contact within < 10 μm
High voltage	30 squares–non-Vcc substrate contact around nwell structure
	Use double ring and grounded fill pattern around high-risk areas

*Dummy structure arrangement with two identical devices at the perimeter (A-A) and two other devices in the center (B-B).

stress are due to macroscopic, nonscalable geometries (e.g., pads, scribes, seal rings, and package materials), which act as interfaces between the IC circuits with submicrometer dimensions, with no air gaps or other boundaries to divert the stress, and the large-scale outside world. Metals and dielectrics on the semiconductor substrate are subject to stress propagation, due not only to the internal die architecture, but also to the external features required for die handling, bonding, packaging, and applications. Mechanical failures manifest themselves as horizontal and vertical decomposition of the die, known as cracks and delamination, enhanced by temperature, pressure, and humidity, and are screened out by the pressure cooker PCT (pressure cooker test) test. We review them next based on the groups of materials.

2.2.2.1 Metals

Reliability issues for metal layers are related to two effects: mechanical decomposition and electromigration (EM). Metals deposited over isolating materials create stress in the underlying layers due to the mismatches of thermal expansion coefficients and difference between the deposition and the operating temperatures (e.g., 400° and 50°C, respectively). As a result, they peel or delaminate (horizontal decomposition) or crack (vertical decomposition) due to the abrupt transitions at terminating surfaces: layer to layer or sharp turns of the geometries on each layer. Of interest to IC construction are a number of metals: Al, Cu, W, Ti, Co, and their alloys. Thin metal layers (i.e., the ones with an aspect ratio of vertical to horizontal dimension <<1 within the area of a simple cell) usually do not present a risk for IC reliability due to good adhesion to the substrate and flexibility. Some films (e.g., amorphous alumina and silica) may show cracking and delamination during the deposition above a critical coating thickness due to intrinsic rather than to thermal stresses.

Cracking and delamination may not occur simultaneously (Figure 2.7). After delamination, the deposited films may curl to scrolls, indicating stress. There is no comprehensive model of the stress gradient in the deposited film to ensure high reliability for all material and geometry combinations in the IC, but there are several methods to reduce the stress magnitude:

- Replace 90° turns by 45° ones to prevent cracking
- Use slots on large areas of metals to prevent EM
- Encapsulate the circuit to prevent moisture penetration

The optimal length of the 45° edges needs to be simulated for the different materials and thicknesses. The slots, which reduce EM, may introduce stress due to the discontinuity of the material and should not be lined up with their edges requiring 45° beveling. To prevent top metal delamination, die corners may be anchored to the substrate.

One should note that beveling introduces additional database address nodes, which, if used too extensively, may extend mask writing time and compromise the quality of pattern transfer from design to wafer.

FIGURE 2.7
Cracking and delamination of ILD and metal layers. (From Y. T. He et al. 2006. 7th International Conference on Electronics Packaging Technology Conference. *IEEE*. With permission.)

Beveling and slotting features may not scale down with the dimensions of other features within the die, driven by the subsequent technology nodes, because they depend on the physical property thicknesses of the deposited materials and the die size, rather than on the horizontal dimensions of the circuit geometries.

A second group of DfR issues in IC metals is due to their sensitivity to EM. EM, known for over 100 years [4], became of interest for IC technologies in 1966, when interconnects were about 10 μm wide. High current densities in metal busses promote diffusive movement of the metal ions due to the momentum transfer from the conducting electrons. Over time, the ions and atoms move from their original positions, causing a void (open circuit), pile up, and diffuse toward other nearby conductors, creating a short circuit called a hillock or whisker failure. Grain boundary diffusion dominates in aluminum wires, whereas surface diffusion is dominant in copper interconnects. In a uniform metal or semiconductor crystalline structure, there is hardly any momentum transfer as the conduction electrons moving through it would experience no collisions with the metal ions. Only lattice defects and thermal vibration of the atoms cause electrons to collide with them and to scatter. The metal bonds at the grain boundaries and material interfaces are weaker. Once the electron wind has reached a certain strength, atoms become separated and transported in the direction of the current, mostly along grain boundaries. Therefore, high current densities ($>J_0$) and the joule heating are needed to cause the atoms to vibrate further from their lattice positions. Jim Black of Motorola formulated an empirical life-span equation under external heating and increased current density that became the backbone of EM modeling [5]:

$$MTTF_{EM} = A_{EM} \cdot \frac{(J - J_0)^{-n}}{T^m} \cdot \exp\left(\frac{E_{aEM}}{kT}\right) \qquad (2.3)$$

Variable	Explanation
$MTTF_{EM}$	Mean time to failure
A_{EM}	Empirical constant depending on the cross-sectional area
J	Current density through the interconnects, $J = \dfrac{CV_{dd}}{WH} \cdot f \cdot P_r$
J_0	Critical current density, below which EM will not happen
E_{aEM}	EM activation energy: grain boundary diffusion: 0.5–0.6 eV (for Al and Al + low percentage Si), 0.7–0.9 eV (for Al + Al alloys doped with a low percentage of Cu).
	Surface diffusion: 0.9 eV (Cu)
k	Boltzmann's constant
T	Temperature (kelvin)
n, m	Material and failure mode-dependent scaling factors
C	Integrated parasitic capacitance: to ground and perimeter
W and H	Width and thickness of a metal line

Variable	Explanation
f	Clock frequency
P_r	Probability that the line toggles in a clock cycle
	EM model dependence
$n = 2, m = 0$	Original Black's equation is not justifiable physically: extracted E_{aEM} is inaccurate but has low impact on lifetime calculations
$n = 2, m = 2$	Shatzkes and Lloyd model for nucleation of damage from vacancy concentration or tensile stress
$n = 1.1, m = 0$	For Cu interconnects

As reflected in Equation 2.3, EM characteristics depend not only on the composition of the metal alloy, the dimensions of the conductor, the crystallographic orientation of the grains of metal, and the deposition parameters, but also on the DC (direct current) or AC (alternating current) waveforms. The EM theory linked the electrostatic force with the momentum exchange to explain the evolution of atom concentration in an interconnect segment. Because $J \gg J_0$ for narrow, metal lines in sub-micrometer ICs, J_0 can be neglected and Equation 2.3 is simplified to

$$MTTF_{EM} = A_{EM} \cdot (J \cdot T)^{-2} \cdot T^m \cdot \exp\left(\frac{E_{aEM}}{kT}\right) \tag{2.4}$$

Current density J and (less so) temperature T are the deciding factors in EM. At prelayout, J can be estimated based on the probability factor P_r, which represents the activity frequency of a particular wire. In low-voltage, high-power MOSFETs, lateral current through the source contact can reach critical densities. The time to failure at test can be extrapolated to the expected life span of the device under real conditions (high-temperature operating life, HTOL).

IC scaling increased the probability of EM failures due to high power and current density. (Scaling by factor k increases power density by k and current density by k^2). For the lower-level interconnect layers of highest layout density (e.g., metal 1), the preferred conductor is Al due to its good adherence to substrate, high conductivity, and formation of ohmic contacts with silicon. Its high susceptibility to EM is reduced by about 50 times by adding 2–4% Cu, which segregates at grain boundaries, inhibiting the diffusion of Al atoms. Copper itself withstands approximately five times more current density than aluminum at similar EM reliability (due to the different E_A values). For Cu, EM can be further improved by alloying it with 1% of palladium, which, similar to Cu in Al, inhibits diffusion of Cu atoms along grain boundaries. For small grains, more grain boundaries increase the likelihood of EM.

While a wider wire results in smaller current density, which should suppress EM, wire width reduction to below the average grain size of the wire material actually increases the resistance to EM as well. Narrow wires exhibit a bamboo structure, with grain boundaries perpendicular to the width of the

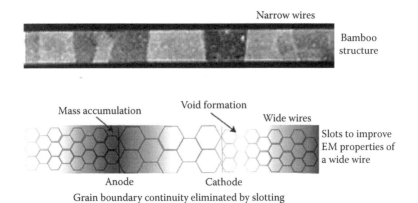

FIGURE 2.8
Bamboo structure preventing electromigration due to the mass transport along the wire. (From A. G. Domenicucci et al. 1996. *Journal of Applied* Physics, 80, p. 4952. With permission.)

wire and to the current flow (Figure 2.8). The boundary diffusion is disabled, and material transport is reduced. However, the wire width smaller than the grain size is usually too narrow to conduct large currents due to the joule heating. As a DfR remedy, rectangular slots are cut along the wide wires such that the widths of the individual metal lines between the slots lie within the area of a bamboo structure. The cost of this remedy is that the resulting total width of all the metal lines has to increase to meet power requirements, also impacting their capacitance. The 90° corner bends must be avoided because current density in such bends is significantly higher than that in oblique angles (135°).

For higher-level metals, EM reliability concerns caused copper and low-*k* intermetal dielectrics to replace Al and SiO_2 at the routing levels. As a result, when designed to CBC principles, ICs should not fail due to EM. Layout rules should correct EM problems readily at the transistor level. Earlier, where proper care was not taken, there have been EM-related product failures. In the late 1980s, the desktop drives of Western Digital (WD) suffered widespread failures 12–18 months into field usage due to an EM rule problem in a third-party supplier's IC controller. WD corrected the flaw, but not before damage to its reputation.

For sub-100-nm ICs, EM in Al, Cu, and W lines or contacts first causes intermittent failures and eventually leads to the loss of circuit connections. The first EM-related glitches are indistinguishable from other failure mechanisms (e.g., electrostatic discharge [ESD]), but at a later stage, EM failures are identified as interconnect erosion-leaving marks visible with an electron microscope.

The impact of grain structure on EM failures of Cu interconnects was investigated by inducing local grain growth by scanned laser annealing (SLA). While Al interconnects show orders of magnitude lifetime modulation for the bamboo grain structures with lengths up to 10 times the interconnect width, no significant differences in the failure rates were found for

Cu, showing that interlevel diffusion barrier layers promote EM by mechanisms faster than grain boundary self-diffusion.

A stress gradient due to atoms driven out of the cathode end of the conductor and accumulating at the anode end modulates not only the wire width but also the atomic density along it. When the wires are shorter than a value called the Blech length (about 10 μm), a mechanical stress causes reverse material migration to compensate the EM mass transport toward the anode. As this may result in a false reliability reading, JEDEC (Joint Electron Device Engineering Council) recommended a design-for-testability (DfT) rule for metal line length for test structures to be much larger than the Blech length (e.g., 400 μm).

Typical current density at which EM occurs in Cu or Al interconnects is over 1,000 A/cm². As the ampacity of W is lower than that of Al or Cu due to its higher resistivity, the current flow must be distributed over multiple W contacts or vias as evenly as possible, especially for power RF MOSFETs optimized for EM (Figure 2.9).

In fact, DfR of RF power MOSFETs requires a compromise between device area, its EM performance, and power dissipation, subject also to the matching restrictions related to macro- and mesoscopic fabrication effects (loading, chemical-mechanical polishing [CMP]).

At the product level, EM for solder joints (SnPb or SnAgCu lead free) occurs at lower current densities (still, over 1,000 A/cm²). The atoms pile up at the anode, and voids are generated at the cathode, with densities depending on

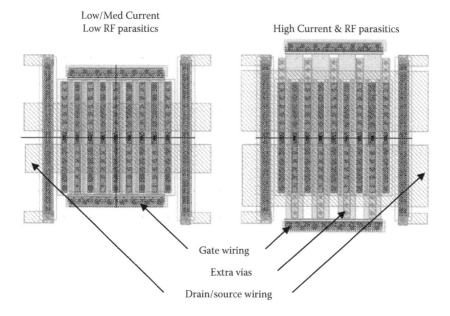

FIGURE 2.9
RF MOSFET with contact density optimized for electromigration.

TABLE 2.10

EM Material-Dependent Properties for MOSFETs

Layer	Material	EM Properties
Substrate	Si, Ge, GaAs, SiGe	No EM issues, single crystal No mass transport due to current flow
Gate	Any	No EM issues, not a current-carrying connector
Contact, via	W Al Cu	Poor reliability of W/Al interface due to atom flux divergence Improved reliability but requires contact enclosures and large CDs
Metal 1	Al	Trade-off: bamboo structure vs. ampacity (slotting helps prevent EM)
Upper metals	Cu	Best reliability but complex process

the back stress. Due to the current crowding, voids form first at the corner of the solder joint and then extend until failure. EM also influences formation of intermetallic compounds.

In summary, DfM for EM in ICs requires detailed knowledge of material properties and models of their physical behavior (Table 2.10). Supporting these models by electronic design automation (EDA) tools at the system level by lifetime DC simulations with SPICE is not a common methodology yet. Existing DfM guidelines for device or block level include

- slotting (for Al only)
- having more contact via area than metal connections
- avoiding metal transitions
- watching out for Blech effect
- using avoidance awareness at small area penalty
- having current extraction at the device level first, then globally

2.2.2.2 Dielectrics

In contrast to the polycrystalline or metal structures, for which reliability problems are related to the grain boundaries, reliability limitations of dielectrics present in IC manufacturing (SiO_2, Si_3N_4, Hf and Ta oxides [6], whether grown thermally or by deposition) are related to their amorphous structure (Table 2.11).

A perfect dielectric should be a material free of charge (i.e., not able to trap electrons or holes) in its bulk or at its interfaces with silicon and gate and nonconductive in the temperature range of application required by the product (i.e., have a large energy gap). For silicon devices, the native dielectric, which goes the furthest toward meeting these requirements, is SiO_2; its low-defect interface to silicon is an all-important feature for MOSFETs. Other

TABLE 2.11

Defects in MOS Dielectric Layers

Dielectric	Purpose	Charge Status during Operation	Defects	Growth Mechanism
SiO_2	Passivate Si surface	Positive	Oxide traps	Thermal growth
Si_3N_4	Improve SiO_2 charge trapping reliability	Negative	Interface traps	Thermal growth or deposition
Hf, Ta oxides	Increase dielectric permittivity	None (conductive)	High leakage	Deposition

materials with custom properties could supplement or replace it (e.g., due to their high dielectric permittivity [Hf oxides] or low diffusivity of dopant species [Si_3N_4 for boron]). However, the charge status of the dielectric during device lifetime needs to be carefully monitored due to its consequential susceptibility to wearout and parametric degradation.

IC topography presents challenges to the mechanical integrity of dielectric layers. If deposited over an uneven surface of conducting layers defined by subtractive etch (active, poly, subtractive metals), interlayer dielectrics (ILD) may be prone to the formation of keyholes, which need extra processing effort (i.e., higher process cost) to control. But, the key physical reliability problems of the dielectrics are due to the mismatches in thermal expansion coefficients among the different conducting and insulating materials in the IC structure (Figure 2.10). Because strong adhesion prevents delamination, thick oxides, which would tend to peel off, and also must not be used in ICs.

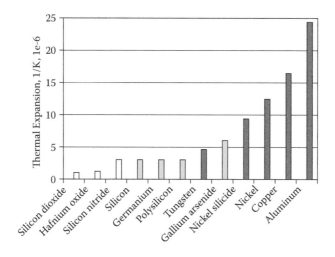

FIGURE 2.10
Material-dependent thermal expansion coefficients in metal oxide semiconductor (MOS) technologies. Dielectrics = white bars; semiconductors = gray bars; and metals = dark bars.

One should note that design and DfM rules pertaining to mechanical reliability are not subject to shrinking e.g., by Moore's law: They do not scale down with technology advances, even though Young's modulus of the material, which, depending on the bond saturation on layer surfaces, determines the potential for cracking and may change with thickness. Therefore, in general, the cost of the die area required to ensure mechanical reliability would increase with every generation of devices. Although mechanical stress at the edge of the die and the dissipated IC power modulating it may not decrease from one technology node to the next, cost considerations may require custom, product-based stress rules depending on the IC form factor and function. At the same time, reliability testing is becoming more complex as IC applications diversify, to cover the many phases of packaging processes such as handling, sawing and singulating, bonding, and molding at elevated temperature. The custom, experimental rules for mechanical reliability for each type of packaging process need significant resources to develop. In this situation it is not uncommon to propose ad hoc, containment layout options to avoid known failure modes in a reactive approach, sometimes without full understanding of the underlying physics. Ex-post simulation and data calibration on one type of product may not be a reliable predictor for another type. A conservative approach is to use a full set of legacy, containment rules for all products and packages unless product qualification helps select the most relevant ones.

The key DfM guideline for IC reliability inside the package is to space the active circuitry away from the macroscopic world represented by pads and die edges. To maximize the gross die per wafer (GDPW) count but prevent delamination, a minimum scribe width rule of thumb may be specified as at least two times the width of the kerf (Figure 2.11) [7]. The corners of the die

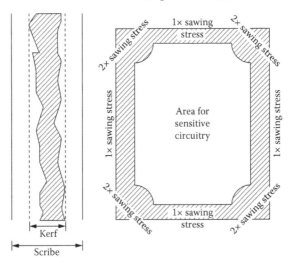

FIGURE 2.11
Regions of die area preferred for placement of sensitive circuitry and IO devices.

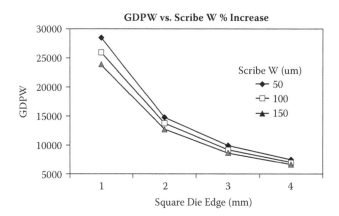

FIGURE 2.12
Impact of scribe size (μm) on reduction of GDPW depending on the size of a square die.

(i.e., intersections of two die edges) are exposed to at least twice the amount of singulation stress, as compared to the single edge, and should not contain any sensitive devices within a predefined, critical radius. Because mechanical stress impacts carrier mobility in transistor channels, active devices must not be placed near die edges (i.e., in the predefined "forbidden areas").

The number of GDPW depends on how conservative are the scribe rules and forbidden areas. For example, for a die size of 1,000 × 1,000 μm, scribe reduction from 100 to 50 μm could bring in several percent GDPW increase (Figure 2.12), which, for a total product revenue of $100 million, translates into several million dollars, which could be spent on improving scribe efficiency (Figure 2.13). One should note that such efficiency requires not only advanced sawing equipment but also carefully chosen scribe content with electrically or physically testable monitors and testing equipment, including precision probe cards. Reducing scribe area requires highly condensed test structures for parametric data extraction, which translates into reduced information content for PY verification. To support new device models, one may need extra scribe area at the expense of GDPW or online scribe-based testing replaced with end-of-line product testing, usually at a higher cost. An alternative is to run a dedicated test vehicle hosting modeling structures. RoI calculations should identify the optimal approach from the scheduling, resourcing, silicon use, and time-to-solution expectations.

The next group of DfR concerns related to the dielectrics is the electrical stress [8]. Because there is no useful or significant current flow through the oxides compared to transistor channels and interconnects, their reliability is related to the impact of leakage and charge trapping (Table 2.12).

Electrical stress in dielectrics may be induced by the nominal voltages (e.g., power supply, VDD) applied for a long time or by elevated voltages, exceeding VDD (also called electrical overstress, EOS) for short or very short times (e.g., due to power glitches) (Table 2.13).

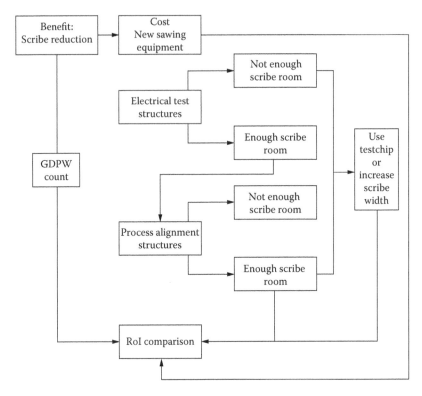

FIGURE 2.13
Process flow to improve scribe line efficiency.

The typical, 10-year gate oxide integrity (GOI) lifetime criterion for active devices is to ensure that no carriers can penetrate the oxide in numbers large enough to cause wearout, for the specified power supply voltage and operating time, even if the carriers permitted to enter the dielectric are deliberately going to change its charge state (e.g., program the memory). Oxide defects due to charge trapping are caused by:

- insulating nature of the dielectric resulting in uncompensated charge centers, which may later decay to neutralized structural defects
- carriers moving across the dielectric, leaving behind a defect trail in the form of conductive paths

The dielectric partially is protected from charge penetration by the energy barriers, usually much higher than the average energy of charge carriers under normal MOSFET operating conditions. However, those barriers on the band diagram of the MOS system (Figure 2.14) may still not be sufficient to prevent charge tunneling through a very thin oxide layer. Therefore, applying

TABLE 2.12

Gate Oxide Integrity (GOI) Degradation

GOI Damage Type	Biased Terminal	Bias Range	Temperature Impact	Failure Mode
TDDB (time-dependent dielectric breakdown)	VG (gate voltage)	$E_{ox} >> 2$ MV/cm	Increases with temperature due to lowered energy barrier	Catastrophic breakdown
HCI (hot carrier injection)	VG, VDS (gate and drain source voltage)	MOSFET in saturation	Decreases with temperature due to detrapping	Parametric degradation due to charging (oxide and interface traps)
NBTI (negative bias temperature instability)	VG (gate voltage)	$2 < E_{ox} < 6$ MV/cm	$>100°C$, increases with temperature	Parametric degradation due to charging (interface traps)

power supply voltage across or along gate oxide over an extended period of time may lead to its wearout depending on the stress magnitude, time, and temperature. The power supply is kept as high as possible because the sub-threshold current (i.e., transistor drain-source leakage) does not scale down with device dimensions and the overdrive ($V_G - V_T$) governing the ON current should not be compromised to ensure high operating frequency and low noise. This gives rise to reliability concerns for devices with gate oxides as thin as several atomic monolayers.

DfR issues in MOSFETs depend on the distribution of the stressing voltage and temperature. GOI reliability challenge due to high gate voltage alone is the time-dependent dielectric breakdown (TDDB). Oxides below 2.5 nm would not be able to sustain their typical operating voltages for the entire desired life of the product [9]. Two models, field driven (E model) and current driven ($1/E$ model), have been proposed to predict the TDDB device life-time [10]. Temperature dependence of time to breakdown for such ultrathin oxides is due to the non-Arrhenius temperature acceleration (for oxides < 6 nm) and increase of voltage acceleration with decreasing voltage [11]. The TDDB

TABLE 2.13

Typical Electrical Stress Conditions in Gate Dielectrics

Voltage (VDD) (% of Power Supply)	Time	Examples
50–100%	Up to 10 years	HCI, NBTI, TDDB
100–200% and more	1 μs (or less) to 1 day	ESD

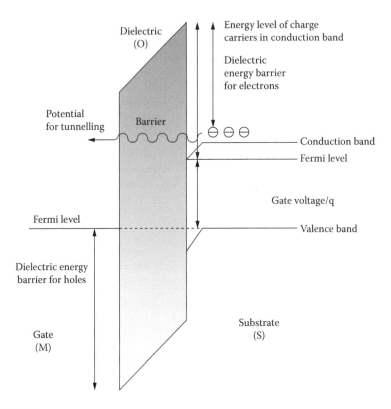

FIGURE 2.14
Energy band diagram of MOS structure showing carrier penetration through the oxide resulting in reliability degradation.

lifetime is proportional to the total gate oxide surface area, due to oxide breakdown being the weakest link [12]:

$$MTTF_{TDDB} = A_{TDDB} \frac{A_G}{(V_{gs})^{\alpha-\beta T}} \exp\left(\frac{X}{T} + \frac{Y}{T^2} \right) \qquad (2.5)$$

Parameter	Explanation
V_{gs}	Gate voltage
T	Temperature
A_{TDDB}	Empirical constant
α, β, X, Y	Fitting parameters
A_G	MOSFET gate oxide area

The impact of oxide defects on TDDB can be interpreted as local oxide thinning, which helps correlate the gate area, pinhole density, yield, and operating voltage with tunneling current as an exponential function of oxide thickness

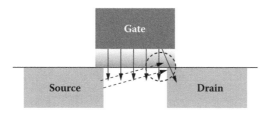

FIGURE 2.15
Electric fields in MOSFET, vertical (solid arrows) and lateral (dashed arrows), giving rise to charge injection into the oxide.

[12,13]. In contrast to the nominal operating conditions, local voltage overstress would be equivalent to hot spot formation and subject to transient modeling.

A process transferred between manufacturing lines with different oxide defect densities would require DRs linking these densities with oxide area and yield. This may be important when adding large-area gate oxide decoupling capacitors to improve on-die voltage distribution and prevent the RF IR (voltage drop due to current flow (I) through a resistor (R)) drop may add pinhole defects and create VDD/VSS (drain source voltage) shorts. Therefore, the number of the gate oxide capacitors should be restricted by DRs, depending on voltage distribution within the power domain.

In summary, DfR for TDDB would depend on the RoI of design and process optimization to meet reliability criteria (FIT) for the product and its manufacturing site.

The next group of gate dielectric reliability issues is related to the lateral field in the MOSFET channel (Figure 2.15), which leads to hot carrier injection (HCI) [14]. MOSFET drive current causes impact ionization near the drain end of the channel due to a high electric field. Charge carriers gain sufficient energy to be injected into the gate oxide, where they create interface traps and fixed-charge, degrading MOSFET threshold voltage, transconductance, mobility, and saturation current. HCI is a significant reliability concern due to the shrinking channel lengths and cannot be eliminated using advanced device architectures (e.g., FINFETs [MOSFETS with channels wrapped by a thin silicon "fin"]). HCI lifetime calculations need to account for voltage acceleration with little effect of temperature. Peak substrate current for n-channel MOSFETs due to impact ionization and peak gate current for p-channel MOSFETs based on the lucky electron model (primarily at low voltages [15]), correlate the device HCI lifetime to only one operating parameter used in SPICE simulation, for a small range of gate voltages near the maximum substrate current (i.e., the drain voltage) [16]:

$$MTTF_{HCD} = A_{HCD} \cdot \exp(\theta/V_{ds})$$ (2.6)

Parameter	Explanation
A_{HCD} and θ	Constants determined from life testing
V_{ds}	Drain source voltage

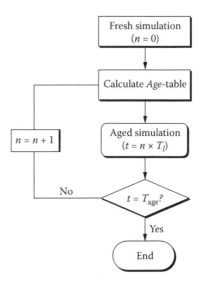

FIGURE 2.16
SPICE-based MTTF and FIT simulations based on oxide wearout models and T_1-elementary time intervals. (From Xiaojun Li et al. 2005. Proceeding of the sixth International Symposium on quality Electronic Design. *IEEE*. With permission.)

This equation supports two methods of reliability simulation [17]:

- Degradation-based model
- Failure-rate-based model

Because device parameters degrade with stress time, it is important to understand how much useful life is left for a product, depending on the time-until-stress date. This can be estimated from the simulated age of the device $Age(\tau)$ (Figure 2.16).

For the degradation-based HCI model, the equation describing the age of the circuit determines its parametric degradation depending on stress conditions:

$$Age(\tau) = \int_{t=0}^{t=\tau} \left[\frac{I_{sub}}{I_{ds}} \right]^m \frac{I_{ds}}{WH} dt \qquad (2.7)$$

Parameter	Explanation
τ	Stress time
m, H	Technology-dependent parameters
W	Channel width
I_{ds}	Drain current (for p-channel MOS, I_{gate} is used)

The $Age(\tau)$ approach, based on static parameters, is used to determine DC lifetime. However, the devices do not stay continuously biased at maximum operating voltages. Therefore, the lifetimes of individual devices need to be calculated based on the actual operating conditions. As a rule of thumb, the nominal DC lifetime can be multiplied by a factor of 10 to reflect time-dependent bias. But, for many product applications, it is critical to extract the age distribution of all MOSFETs on a chip and use geometry derating to ensure reliability of the critical ones.

The oxide failure mechanism related to both high gate voltage and operating temperature, the negative bias temperature instability (NBTI), mainly occurs in pMOS devices stressed with negative gate voltage, at 100–250°C and an oxide electric field between 2 and 6 MV/cm [18]. Thermally activated holes trapped between the silicon dioxide and the substrate gain energy to disassociate the interface/oxide defects near the gate edge at the lightly doped drain (LDD) regions with higher hole concentrations and locally increase the threshold voltage, thereby reducing drain current and off-state leakage. NBTI damage has become more severe than HCI damage for ultra-thin oxides at low electric fields [19], where oxide nitridation prevents boron from penetrating through the gate. The NBTI-related variation in threshold voltage is due to the increase in positive fixed charge ΔN_f and donor type interface traps in the lower half of the silicon bandgap ΔN_{it} [20]:

$$V_{th} \propto \frac{q}{C_{ox}} (N_f + N_{it}) \qquad (2.8)$$

Parameter	Explanation
ΔV_{th}	Threshold voltage shift
q	Elementary change
C_{ox}	Oxide capacitance
N_{it}	Interface trap density $\propto E_{ox}^m t^n \dfrac{1}{T_{ox}} \exp\left(-\dfrac{E_a}{kT} \right)$
T	Temperature
E_{ox}	Oxide field
T_{ox}	Oxide thickness
m,n	Fitting parameters, 1.5–3.0
E_a	Activation energy 0.15–0.325 eV
t	Stress time
N_f	Fixed charge density $\propto E_{ox}^m t^n \exp\left(-\dfrac{E_a}{kT} \right)$, independent of T_{ox} negligible compared to N_{it}

Oxide electric field E_{ox} due to the gate voltage V_{gs} (for p + poly-Si gate pMOS-FETs, $E_{ox} \approx V_{gs}/T_{ox}$ [21]), creates a V_{th} shift of

$$V_{th} \propto V_{gs}^m t^n \exp\left(-\frac{E_a}{kT} \right) \qquad (2.9)$$

The NBTI lifetime or MTTF defined as the time for the threshold voltage shift to reach a predetermined value (e.g., 0.1 V) can be modeled as

$$MTTF_{NBTI} = A_{NBTI} \left(\frac{1}{V_{gs}} \right)^{\gamma} \exp\left(\frac{E_{aNBTI}}{kT} \right) \qquad (2.10)$$

Parameter	Explanation
A_{NBTI}	Process-related constant
E_{aNBTI}	Activation energy 0.9–1.2 eV
γ	Voltage acceleration factor 6–8

Reliability simulations (SPICE) are required to avoid unrealistically expensive testing. To experimentally verify a FIT rate of 10–100 would require over 10^7 device-hours of testing at a 60% confidence level, with no failures reported [17]. The effectiveness of the voltage and temperature acceleration in reliability screening (burn-in) is declining for a diminished gap between normal operation and overstress. In age simulations, spice parameters of unstressed devices enable to accurately model I_{sub}, corresponding to the largest shifts of saturation current I_{dsat}, threshold voltage V_{th} (for analog devices), or maximum transconductance g_{max}. The stress time resulting in 10% shift of one of these parameters is defined as the device lifetime. In AgeMos extraction [17], the degradation models are fed to SPICE-like simulators with Virtuoso UltraSim and an iterative, accelerated lifetime model tool.

The degradation-based lifetime depends on the combined effects of all GOI failure modes. To calculate FIT (i.e., the failure rate for 10^9 device-hours), one has to assume a constant failure rate for each failure mechanism. For failure modes with time-variant characteristics [21], we can assume an IC with a competing series failure system or assume each failure mechanism to have an exponential lifetime distribution. In this way, each failure rate is treated as a constant. We can then apply the standard sum of failure rates (SOFR) to determine the failure rate of a system from its individual components [22]. An IC comprised of n units can be related to the lifetime of each unit ($MTTF_{ij}$) due to each of its m individual failure mechanisms:

$$MTTF_s = \frac{1}{\displaystyle\sum_{i=1}^{m} \sum_{j=1}^{n} \frac{1}{MTTF_{ij}}} \qquad (2.11)$$

The FIT is related to MTTF for a constant failure rate system:

$$FIT_s = \frac{10^9}{MTTF_s} \qquad (2.12)$$

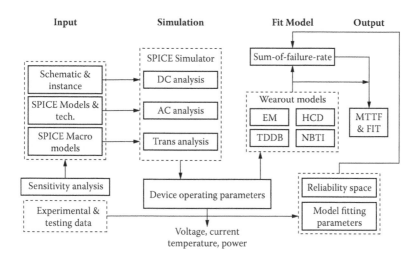

FIGURE 2.17
Failure rate based on age simulation methodology. (From Xiaojun, Li et al. 2005. Proceedings of the Sixth International Symposium on Quality Electronic Design. *IEEE*. With permission.)

Based on the degradation models for EM, HCI, TDDB, and NBTI combined in the SOFR, SPICE reliability simulation based on failure rate requires the same assumptions as the accelerated stress test and considers the impact of device operating parameters (current, voltage, frequency, temperature, and power) on their failure rates, not on their degradation behavior. MTTF and FIT replace the *Age* parameter to model the aging process, based on SPICE simulation of each interconnect and MOSFET in a circuit (Figure 2.17). This method can also be used to define device lifetime under derating conditions, with derating factor D_f defined as the ratio of MTTF for a device at manufacturer-rated operating conditions to the MTTF of a device operating at derated conditions [17]:

$$D_f = \frac{MTTF_{derated}}{MTTF_{rated}}$$ (2.13)

Damage Source	Derating Equation	Derating Factors
EM	$\left(\dfrac{J_0 T_0}{JT}\right)^2 \exp\left(\dfrac{E_{aEM}}{k}\left(\dfrac{1}{T} - \dfrac{1}{T_0}\right)\right)$	J_0, T_0
HCI	$\exp\left(\theta\left(\dfrac{1}{V_{ds}} - \dfrac{1}{V_{ds0}}\right)\right)$	V_{ds0}
TDDB	$\dfrac{(V_{gs0})^{\alpha-\beta T_0}}{(V_{gs})^{\alpha-\beta T}} \exp\left(X\left(\dfrac{1}{T} - \dfrac{1}{T_0}\right) + Y\left(\dfrac{1}{T^2} - \dfrac{1}{T_0^2}\right)\right)$	V_{gs0}, T_0
NBTI	$\left(\dfrac{V_{gs0}}{V_{gs}}\right)^{\gamma} \exp\left(\dfrac{E_{aNBTI}}{k}\left(\dfrac{1}{T} - \dfrac{1}{T_0}\right)\right)$	V_{gs0}, T_0

For EM, HCD, TDDB, and NBTI reliability assurance to be based on derating throughout a specified lifetime, it should be verified first by RoI trade-off calculations against reduced circuit performance.

The MTTF/FIT simulation with Spectre RF can be expanded (e.g., to characterizing HCI and SBD in RF ICs). By combining with the aged model files for LNA (low noise amplifier) and VCO (voltage controlled oscillator) one can prove, for example, that the use of a cascaded structure reduces hot carrier degradation.

One goal of GOI simulations is to find reliability hot spots (i.e., weakest MOSFETs), which as a result of degradation would contribute most significantly to the propagation delay, making the circuit exceed the specified maximum degradation and then fail. A simulation of the reliability critical path (RCP) would then determine the circuit lifetime. If that lifetime is below the required limits causing a loss of the marketable value, local design for reliability, LDfR should be used to improve the reliability but without significantly changing the circuit performance. Limited adjustments to device properties may suffice (e.g., increasing channel length of selected MOSFETS to improve HCI reliability). DfR optimization at minimal scope is required to avoid a chain (domino) scenario, where a longer gate would slow the hot spot circuit and its channel width would need to be increased (Figure 2.18). This in turn would increase the load capacitance and reduce the speed of the preceding gate in the RCP. For complex logic circuits, one would have to maintain the speeds along paths other than RCPs. Redesign may include adding a dual power supply, which requires a dual threshold voltage setup to maintain the

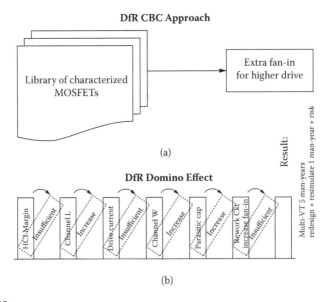

(a)

(b)

FIGURE 2.18
DfR domino effect: (a) object of analysis and requirement; (b) domino (chain) scenario. Area increase, capacitance, and fan-in.

performance, creating other reliability issues, such as ESD failures or leakage. Dedicated methodology may be required to ensure a reliability convergence path that does not have an impact on the project schedule.

2.2.2.3 Semiconductors

Similarly to the metal and oxide layers, semiconductor substrate is subject to physical and electrical stress during device packaging and operation. However, a perfectly crystalline structure of the substrate is able to withstand all kinds of stresses except those leading to wholesale destruction. Unlike for the metal and oxide, reliability concerns for the semiconductor substrate are therefore related to transient effects, which, if not properly screened, may lead to circuit damage. Permanent stress effects, such as mobility variations related to material properties are no longer qualified as poor DfM but as advantages to the device design.

Semiconductor substrate absorbs mechanical stress not only from the overlying oxide and metal but also from the external stress forces, which are not buffered by these layers. For the stress not to change device properties, a cushion layer (e.g., polyimide) may be required, which increases process cost. Alternatively, a dedicated environment for sensitive circuits may be required, such as special shielding architecture at multiple metal levels.

Because mechanical integrity of the devices is guaranteed by the strength of the monocrystalline substrate, the sturdiness of alternative substrates for future ICs, such as organic materials, has to be carefully studied before product introduction. To protect from the impact of the bonding forces, active circuitry has been excluded from under pad areas occupying significant room in the die. Recently, less-sensitive devices have been allowed under pads as a cost reduction measure. To save as much as 10% of die area, circuits under pads (CUPs) include high-current inputs and outputs (I/Os) with large geometries, which are not as sensitive to stress-related changes except for the transient bonding stress. Extensive product qualification is required to ensure CUP reliability [23].

The relative area occupied by the pads depends on the die size and architecture. For example, for a $1,000 \times 1,000$ µm die, a single 50×50 µm pad would occupy 0.25% of it (i.e., 10% for a 40-pad product). Therefore, significant percentage of product development efforts can be committed to utilizing the CUP area.

Electrical integrity of the substrate depends on the types of isolation built into it to prevent a discharge path between the power supply, the I/Os, and the ground. Current flowing through the substrate can create discharge paths between the different I/O pins due to the misbiased junction isolation between the adjacent devices or between any active device and the substrate. While a bias independent dielectric isolation to the substrate (e.g., using silicon-on-insulator wafers) would significantly increase process cost

[24], accidental forward biasing of a normally off diode can create a flood of carriers, changing the bias state of other reverse-biased junctions between V_{cc} and V_{ss}. As a result, latchup is a key electrical transient effect in the semiconductor substrate, depending on a combination of biasing conditions and external sources of electron-hole pairs [25].

An IC coming out of a manufacturing line contains considerably more devices than necessary for its function and reflected in its schematic. These extra devices (transistors, diodes, and capacitors), called the parasitics, result from the IC construction, but should be inconsequential under normal operating conditions. However, exposure to overstress or high-energy radiation may activate them and compromise the correct operation of the product IC. Junction isolation of the individual devices may cause activation of parasitic transistors, even if their current gain is very small ($\beta < 1$). While a considerable stimulus current in the form of long overstress pulses is necessary to activate these devices due to their low transit frequency f_T (typically in the megahertz range) interference in sensitive analog circuits may trigger a latchup and generate a short circuit. In Figure 2.19, the P- and N-doped regions in a CMOS inverting amplifier form source and drain and the cathodes of the clamping diodes are connected to the most negative point in the circuit (e.g., ground [GND]), such that the P-N junctions are blocked. The P-substrate forms the base of a parasitic NPN transistor, with N-doped regions functioning as emitters and the collector formed by the well in which the complementary P-channel transistor is located. The NPN and PNP transistors form a thyristor, with the anode and cathode connected to the supply voltage, while all other points (I/Os) function as the gate. As long as the voltages on the connections stay between the ground and the VCC, the base-emitter diodes are blocking, and device operation

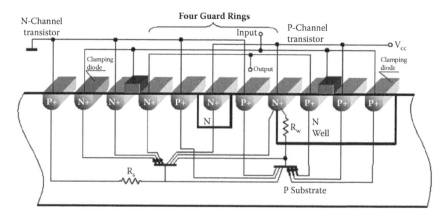

FIGURE 2.19
Parasitic transistors causing latchup in the CMOS circuits and guarding-based protection.

is not compromised. Otherwise, the thyristor can be triggered in various ways:

- By voltages at the input or output of a circuit, more positive than the supply voltage or more negative than the voltage substrate, causing current flow longer than 1 μs (not likely to happen due to line reflections along the connections on circuit boards, but possible for lines of several meters in length, e.g., at the interfaces between the circuit and the outside world).
- By ESD, which, while lasting only a few tens of nanoseconds, can flood the IC with charge carriers, slowly flowing away and creating the potential to trigger the thyristor. A rapid rise of the supply voltage may increase the current in the parasitic transistors, leading to an avalanche process to activate the thyristor.
- By ionizing radiation (even low doses), also creating a flood of charge carriers due to electron-hole pair generation.

A triggered thyristor, which enters a very-low-resistance state, short-circuiting the supply voltage and given rise to a very high current, may either damage the component or have to be switched off only by turning off the supply voltage. The role of the latchup-protection DfM is preventing these events in the schematic or layout of the device. A simple DfM latchup protection is to insert a resistor in series with the supply voltage to the IC to limit the current to a value no longer dangerous to the device or, better, to a value below the holding current of the thyristor if it is switched on accidentally. Other, more space-consuming (costly) DfM option to prevent latchup include adding space between the devices to reduce layout sensitivity. If the potentially conflicting components are placed far away from each other, the current gain of the parasitic transistors related to them is reduced. A superior method of layout-based latchup protection, surrounding the critical circuits with guard rings tied to the positive or negative supply voltage close to the base emitter region to enable charge collection away from the parasitic transistors (Figure 2.19). In all these cases, RoI analysis of the incremental die area or design complexity versus product-dependent reliability would help select the best option for ESD protection, also enabling routing of high-energy discharges, which, by nature, requires volume, but steering clear of engineering solutions that are too big (expensive).

A related parasitic effect for bipolar transistors in analog circuits is called pseudo-latchup. A voltage applied to the inputs of an operational amplifier or comparator, outside the range for common-mode operation, puts the input of the circuit into a state that results in unpredictable behavior (pseudolatchup), although it is unrelated to the thyristor turn on (Figure 2.20). While a defined state of the circuit can be reached again, when the voltage at the inputs returns to the data sheet value, the internal circuitry might become latched. Then, a

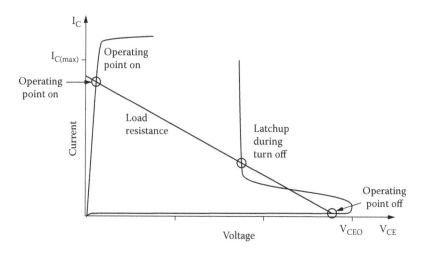

FIGURE 2.20
Output characteristics with load resistor. (From E. Haseloff. 2000. Application Report SLYA014A-Latch-Up, ESD, and Other Phenomena. Texas Instruments, Dallas, TX. With permission.)

reduction of the input voltage no longer results in the correct output voltage, and the power supply must be completely switched off to reactivate the amplifier. Another type of pseudo-latchup occurs for bipolar output transistors at high currents. It is not critical if the transistor is switched on fast, although the breakdown characteristics may be exceeded briefly and the transistor's resistance decreases. When the transistor is switched off, its current flows along the load resistance line in the "off" direction. When the transistor intersects the breakdown characteristics, it "latches up" as it becomes increasingly resistive. A considerable collector current flows for a high collector-emitter voltage, and power dissipation increases the IC temperature, resulting in device aging or destruction. Touching the collector with a probe or capacitive loading may prevent the latching. Latchup DfR-DfM can only be complete when all these parasitic devices are extracted and simulated for their current gains.

Another electrical IC reliability issue related to active devices in the substrate is the electro static discharge (ESD) [26]. The electrostatic energy flow from the bodies or objects into the ICs it touches in the process of their handling and operation can destroy the electronic circuits. The severity of the ESD events depends on parameters of the charging and discharging, such as total charge, speed of charging, and volume of device and its protection circuitry, driven by the correspondence to real-life situations.

Several models have been created to optimize DfR/ESD preventive measures (e.g., by utilizing clamping diodes on the inputs and outputs of logic circuits to limit voltage excursions) (Figure 2.21). The human body model (HBM) [27] simulates a relatively benign situation in which the energy stored in a human body is discharged into an IC by substituting that body with a

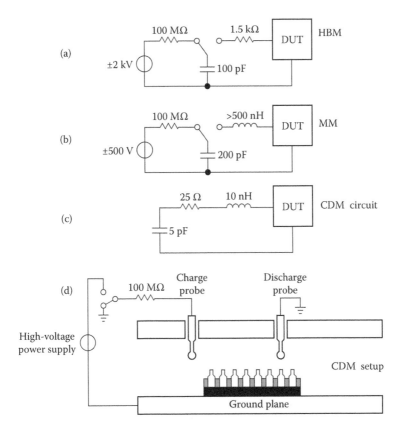

FIGURE 2.21
ESD test circuits corresponding to (a) human body, (b) machine model, (c) charged-device model, and (d) charged-device model test setup. (From E. Haseloff. 2000. Application Report SLYA014A-Latch-up, ESD, and Other Phenomena. Texas Instruments, Dallas, TX. With permission.)

capacitor (C = 100 pF) charged through a high-value resistor to 2,000 V and then discharged through a 1.5-kΩ resistor into the device under test.

A more harsh but realistic HBM model assumes a capacitance of 150 pF, a discharge resistor (R = 330 Ω), and a body charged to 4,000 V, as in a discharge happening in the tips of the fingers [28]. While the energy of about 0.4 µWs dissipated in the actual protection circuit is comparatively small, the rise time of the discharge current controls the area of the conducting region in the protection circuit, which, for the initial 1 ns, can be locally overloaded. The discharges at a slower rate, even from high voltages, may not destroy the circuit if they do not occur directly at the pads of the IC but are separated from them by the length of a conductor with significant inductance.

ESD events in the IC packaging process are simulated by the machine model (MM) [25]. The test circuit with a C = 200 pF capacitor charged to _500 V, then discharged without a series resistor into the device under test, creates a peak current significantly higher than in the HBM. As a realistic mitigation, the inductance of 500 nH of the MM controls the rise time such that the amplitude of the current and the local overload of the protection circuit are reduced. ICs usually cannot be MM tested at 500 V without damage due to the high energy of the discharge.

ESD during IC processing is simulated by the charged-device model (CDM) [25]. The ICs are charged when sliding along plastic transport rails before being discharged when inserted into circuit boards. Because the IC package is small, both capacitances and inductances are low, leading to short current rise times (<200 ps) and resulting in damage due to the partially conducting protection circuit. The device under test (DUT) is placed on its back on a metal plate to ensure the largest possible capacitance of the circuit to the environment, charged with one test probe and discharged with a second test probe. ICs need to survive charging/discharging up to 1,000 V to be assembled without problems. Unfortunately, there is no direct correspondence between the results of the HBM and CDM tests. Therefore, DfM approaches covering these types of ESD failures have to be different, depending on the prevailing risk during the IC handling. As one way of correlation among the different stressing conditions to define common DfM solutions, components that survive the 2,000-V HBM test without damage are assumed to be immune to a 200-V MM test.

Another ESD problem for ICs inside electronic equipment charged cable model (CCM) can be created by touching a cable connector by a user who first walked on a nonconducting floor, holding a plug of a long cable in hand and charging up in the process [25]. The plug is then inserted into the socket, discharging the capacitance of the cable. For a 10-m cable with $C = 1,000$ pF and the charging voltage of 1,000 V, the 500-µWs of energy, which must be tolerated by the IC, are many times larger than assumed in the HBM, MM, and CDM models. The discharge current, determined by the line impedance of the cable (typically 100 S), is about 10 A and flows for a time corresponding to the doubled signal propagation time (i.e., about 100 ns). Luckily, the high inductances of the connector do not allow for steep current-pulse edges, and the protection circuits can conduct over their entire areas, reducing the discharge intensity.

To tolerate high discharge energy levels and to protect more sensitive parts of the IC, DfM for ESD may require input protection circuit consisting of two stages (Figure 2.22):

- Stage 1, a coarse protection, such as separate protective circuits against negative and positive voltages from the outside, to conduct away the higher-energy levels.
- Stage 2 protecting the individual I/Os depending on the current distribution inside the IC.

For the lowest cost of DfM i.e., using only the required protection circuitry, design has to be customized for the IC structure, depending on

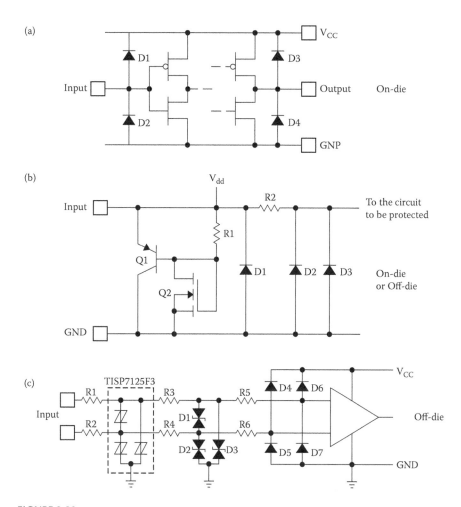

FIGURE 2.22
ESD protection circuits: (a) diode, (b) two stage, (c) multi-stage extreme. (Adapted from E. Haseloff. 2000. Application Report SLYA014A-Latch-up, ESD, and Other Phenomena. Texas Instruments, Dallas, TX. With permission.)

whether the IC contains operational amplifiers, high-resistance input circuits, or interface devices etc. Implementation of protection devices depend also on the limiting properties of diodes. Excessive interfering voltage at the I/Os of the circuit (e.g., a double operational amplifier) can prevent it from functioning correctly. Because silicon or Schottky diodes have too high forward voltage ($V_f = 0.7$ V or 0.4 V, respectively) and germanium diodes, while more suitable ($V_f = 0.3$ V), have a low operating temperature ($T_{max} = 90°C$), the current limiting should be done by feeding the bases of transistors, such that the input voltages stay below -0.2 V. For voltage limiters connected in parallel with the inputs of the ICs, resistors must be inserted to limit the input current (Figure 2.22c). The limiter circuit has a comparatively high input capacitance, which, together with the series input resistance R_s, influences the IC upper-frequency limit. Digital-to-analog (D/A) converters, which contain a capacitor network, contribute additional input capacitance, which acts as a low-pass filter.

DfM optimization of ESD protection circuits has first to consider the low rate of thermal conduction of silicon, at 1 µm/µs, requiring the protection circuit to withstand the total energy before the heat is conducted to the surrounding areas. The CCM model would best approximate the area necessary to withstand this stress. The temperature increase of the protection circuit

$$T = \left(\frac{1}{V \cdot C_s}\right) \cdot E \qquad (2.14)$$

Parameter	Explanation
ΔT	Temperature increase
V	Volume of the protection circuit to be heated
C_s	Thermal capacitance of silicon = 1.89 W/(cm^3× K)
E	Injected energy

should not exceed ΔT = 150 K to ensure that the circuits return to normal operation after cooling. For E = 500 µWs, the minimal volume of the protection circuit is

$$V = \frac{500\,\mu Ws}{150K \cdot 1.89\ Ws/cm^3 K} = 1.76 \cdot 10^{-3}\,mm^3 \qquad (2.15)$$

which, with the injected energy converted into heat in the depletion layer 2 µm thick, makes the necessary area A equal to

$$A = \frac{V}{D} = \frac{1.76 \cdot 10^{-3}\,mm^3}{2\,\mu m} = 0.88\,mm^2 \qquad (2.16)$$

For a total IC area of only a few square millimeters, not only would the cost of the protection circuit be a significant percentage of the total cost of the IC but also such a circuit would increase the input capacitance and induce leakage currents critical for operational amplifiers. As an alternative to the on-die clamps, one can limit the excessive currents and voltages by external protection circuits (Figures 2.22 b,c). These are indispensable, for example, for protecting against lightning strikes during a storm (lightning electromagnetic pulse, LEMP), which induce thousands of volts and hundreds of amperes into nearby conductors.

IC interfaces to the outside world are often unpredictable. Therefore, IC DfM must make the circuit reliable for an extended range of applications. Telecommunication and data transmission installations require special voltage limiters. With three-dimensional (3D) integration of multiple chips, one may add dedicated protection circuits as separate dice to withstand severe conditions for the advanced applications of SoC products.

Extreme environmental hazards require three-stage protection circuits (Figure 2.22c). The first stage consists of the external voltage limiters and drains away currents on the order of several hundreds of amperes. At the second stage, clamping diodes lead off currents in the range of amperes. The third stage consists of the ESD protection circuits inside the IC structure. As a result, the input of the differential amplifiers is protected against both unipolar and differential interference as long as the grounding is properly selected.

Malfunction of digital ICs due to parasitic transistors is much less likely than for analog circuits because relatively higher currents are needed to cause an incorrect logic level, and guard rings drive away parasitic currents at the I/Os. However, reflections at the ends of lines not terminated correctly may cause dangerous voltage overshoots and undershoots.

On-die clamping diodes limiting the voltage excursions (Figure 2.22a) must be able to pass relatively high currents, on the order of 100 mA. This assumes that a switch-on duration of the current is only a maximum of 100 ns, with a duty cycle less than 10% (i.e., a fraction of the period of the valid signals). Good control of the circuit switch-on time makes it possible to lower the area of the guard rings because short pulses (in the range of nanoseconds) are unable to switch on parasitic transistors. A DC test at $I_D = 3$ mA injected into the clamping diodes for a duration of $t_d = 10$ to 20 μs, correlates to the assumed operating conditions of 80 mA/100 ns.

In summary, parasitic devices and currents in ICs cause operational problems in the many unexpected, nonstandard use conditions, depending on the analog or logical circuit application and on the parameters of the interfering signals: frequency, rise time, amplitude, or energy. IC DfM rules should help select and cost optimize the protection architecture, clamping devices, and guard ring area.

Latchup in ICs can also be triggered by radiation (soft-error latchup, SEL) from three sources [25]:

- cosmic rays, absorbed by ICs working at elevated locations
- alpha particles due to the materials used for IC protection (mainly boron)
- thermal neutrons from the environment.

One response to the soft-error rate (SER) transient effects can be based on MfD rather than DfM, by modifying doping profiles using epitaxial layers or implant triple wells. Such solutions bear a high fixed cost compared to the DfM basic approach, such as using circuits with high node capacitance.

The IC design architectural problems to be addressed by DfM also include on-die, within-package, and onboard power distribution. The growing range of IC applications requires DC and AC power domain optimization, e.g., by IR drop simulations, routing rules and best practices, adding decoupling devices, and so on. Power domain issues expand IC DfM into the applications domain, making it necessary to consider the RLC (resistance inductance capacitance) parasitics of the package and printed circuit board (PCB). This group of issues is now emerging also in relation to packaging techniques. A separate DfM discipline will soon be needed to comprehensively cover the die-to-product interaction.

2.3 Design for Test

The precursor to high reliability in the field is high quality at test. For complex ICs such as SoCs, the cost of testing could match the cost of manufacturing due to both the extended test time and equipment complexity [29]. Optimization of the testing process should therefore be considered a long-term investment and divided into four phases: Planning, implementation, execution, and final verification and debug. The CBC approach should enable focusing on cost reduction [30].

A variety of product applications and failure modes makes the trade-off between CBC and comprehensive testing a difficult one. While the earlier a test is applied in the product manufacturing cycle, the lower is its cost, and the most cost-effective testing should be considered the parametric testing online supported by a limited set of scribe line structures, the scope of SoC design may be too large to rely on CBC alone for high testing efficiency.

Following the rule of 10, the cost of testing increases from parametric to sort to class. Simultaneously, the size of the tested sample often decreases as it is becoming more difficult (expensive) to test a large quantity of devices by complex tests. Efforts are concentrated on eliminating as many testing procedures as possible, which also creates a significant need for DfT.

Design for test (also known as design for testability, DfT) is a name for design techniques that add testability features to an IC to make it easier to develop and apply manufacturing tests to it [31]. Test programs run the automatic test equipment (ATE) or the assembled system to find defects (if the test fails) and log diagnostic information about their nature. Automatic test pattern generation (ATPG) is much easier if DfT rules and best practices are implemented.

DfT for electronic applications has been known since the 1940s and usually for DfT associated with product design with the primary goal of enabling easy access to the selected internal circuit elements. Design enhancements can be architectural (e.g., adding active circuits to facilitate controllability/observability, such as adding a multiplexer into a net) or physical (e.g., adding a probe point to a net). DfT also deals with such electromechanical characteristics as the size, shape, and spacing of probe points or high-impedance drivers attached to the probed nets to mitigate the risk of damage from back-driving [32].

DfT and functional IC testing to specifications are based on different premises. IC DfT is supported by structural tests to make sure that the circuit has been assembled correctly from building blocks and to stipulate that if the netlist is correct and structural testing has confirmed correct assembly, the IC should be functioning correctly.

Structural test generation only deals with primary I/Os and physical test points, so the embedded gates must be manipulated through the layers of logic with exponential state sequencing. DfT removes the state transition sequences observing the status of the internal circuit elements, to perform the same function as a less-expensive alternative to upgrading the test equipment. While DfT has to match advances in I/O count, size, placement (spacing), speed, internal circuit complexity, power and thermal control, and so on, the test compression ensures that tester times stay within bounds dictated by the cost target for three basic test types (Table 2.14).

One key approach to DfT is scan design, which propagates test patterns from the IC inputs to the internal circuits under test (CUTs). The registers (flip-flops or latches) are connected in chains to gain access to internal nodes of the IC while test patterns are shifted in, and clock signals are pulsed to test the circuit during the "capture cycles." The results are then shifted out of the chip output pins and compared against the expected "good machine" results. Large scan vector sets that take a long time and large memory are decompressed at chip input and compressed at the test output, allowing for large time gains as the test vectors usually need to set and examine a fraction of the scan chain bits [33].

In addition to the "go-no-go" testing, scan chains can also be used for debugging. The IC function is assured, the clock is randomly stopped, and the IC reconfigured into test mode. The internal state of the IC can then be dumped out using the scan chains. Another use of scan to aid debugging consists of scanning in an initial state to all memory elements and then performing a system debug to reduce the time of setting the system to a known state. This use of scan chains along with the clock control circuits is a related subdiscipline of logic design called design for debug (debugability).

Electrical and physical design have to respond to the needs of testing tools and procedures, based on DfT RoI analysis. For example, logic built-in self-test

TABLE 2.14

Basic Types of IC Tests

Test Type	Purpose
Volume	Identify systematic failures
Fast	Laser repair: quickly find the failing unit and create a repair work order
FA friendly	Simplify data collection and diagnostics for low-cost failure analysis (FA)

(BIST) does not work well with random-defect-resistant designs. Design conversion for full-scan ATPG would have an impact on cost and schedule [34].

DfT closure [35] i.e., rapid, predictable, and repeatable approach to satisfy design and test requirements has to converge on all constraints (function, timing, area, power, reliability, routability, and testability) and must start very early in the design process. A register-transfer level (RTL) designer has to deliver synthesizable code in Verilog or VHDL with test vectors of high quality. DfT tools must also have no impact on timing closure flows down the line.

DfT analysis must therefore begin at RTL and proceed hierarchically as follows:

- Physical synthesis would implement DfT architecture with constraint optimization and awareness of test mode behavior and requirements (clock domains, capture groups, optimal scan routing, etc.).
- Creation, verification, and management of test design data would be automated from all design database attributes, initialization sequences, and protocol information and vector files.
- No design DfT handoffs between discrete processes are acceptable, such as synthesis and scan insertion causing iteration loops due to the problems later in the design flow, where a designer can unknowingly break DfT rules.

DfT, just as each design deliverable, must be self-contained but requires full understanding of the IC application as an implementation of design database. For example, timing engines, which forward-annotate timing constraints from high-level design to physical synthesis, can reduce iterations in complex SoC designs. DfT needs to be closed in parallel to ensure these SoCs are testable without iteration loops between synthesis and test activities.

2.3.1 DfT Techniques

The optimal DfT would make all netlist nodes controllable and observable. The state of sequential registers can be controlled from the input boundary of the chip, and design response can be captured and observed at output boundary. Scan is the most popular technique to increase testability compared to functional testing and stuck-at testing (i.e., the traditional testing methods). With functional testing, the tester applies a sequence of input data and detects the resulting sequence of output data compared against the expected behavior of the device, as in the target application. However, this testing has limited ability to test the internal nodes. Using ATPG, one can test a much larger number of internal faults than with functional testing alone. ATPG sets all nodes of the circuit to both 0 and 1 and propagates the defects to nodes where they can be detected.

Full-scan methodologies involve replacing all sequential registers of the DUT with scan equivalents connected serially in one or more scan chains [36]. Test stimuli applied to internal nodes of the design by shifting data in through these chains allow for design response by capturing and shifting the data out. This technique yields very high levels of controllability for high manufacturing fault coverage. Scan methodologies present a combinational view of the sequential behavior of the DUT so that combinational test pattern generation techniques can be used to generate high-quality test sets. However, due to the serialization of test data, scan involves a trade-off between test application time and design resources, with shorter scan chains requiring less test time but more design-level access ports and vice versa. Also, replacing registers with scan equivalents increases the die area. Partial-scan methodologies selecting a subset of registers to include in scan chains may be used as a trade-off based on topological analysis, testability measures, and so on.

Test points improve testability in full scan, adding controllability of register clock/set/reset signals (Figure 2.23). While algorithms based on testability or topological analysis help identify netlist nodes where test points can be inserted to reduce test data volume, designers are very careful when inserting them because of the logic overhead. For SoC, designers integrate a number of off-the-shelf modules (cores) provided by different vendors. Typically, these cores are supplied with associated test pattern sets. It then becomes the integrator's job to assemble the test patterns of various cores into a chip-level test program so that each of them can be tested from the design boundary by reusing the supplied test pattern sets. The cores must be encapsulated so that their interface signals are accessible from the chip interface, typically "wrapping" their ports through a chain of sequential registers. The core test

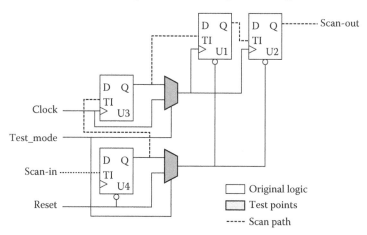

FIGURE 2.23
Test points on clock and asynchronous signals, adding logic tests to the signal flow. (Reprinted with permission from Synopsys, Inc.)

patterns must also be ported to the chip-level test program, and a test schedule must optimize the testing sequence. Cost reduction is achieved by EDA techniques to reduce test vector size, data volume, and speed up test generation, and by ATE companies driven by IC manufacturers to supply low-cost testers and easy transfer of test vectors from EDA tools.

One important DfT methodology is BIST, which creates test patterns "on chip" through a pseudorandom pattern generator, applying data through scan chains and compacting the design response to a signature that can be matched to the expected response [37]. Since the test stimuli are generated on chip and the response data are compacted, BIST provides significant savings of test data volume, application time, and equipment cost. However, it may lead to poor test quality for designs resistant to random-defect-resistant designs and to increased die area. To remove the conflict between the test quality and chip area, test compression is becoming mandatory for additional at-speed and bridging tests while supporting existing test budgets. Adaptive scan technology [35] delivers 10–50× test compression within design synthesis, optimizing the traditional scan chains into smaller segments for time savings, while the adaptive scan load decompressor and unload compressor reduce the amount of required test data needed, lowering ATE memory requirements (Figure 2.24).

Other test techniques include data encoding and boundary scanning. Reordering scan chains and selecting nearest neighbors based on wire length help address area overhead and timing by minimizing buffer insertion for a fixed hold time [36].

Reordering scan chains in the place and route tools requires a list of "stub" chains from DfT synthesis. Their boundaries can be an I/O port, a lockup latch, or a multiplexer. The implementation of a physical scan chain has to ensure the ability to check integrity for optimization. DfT synthesis

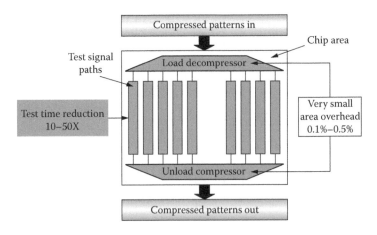

FIGURE 2.24
Adaptive scan technology. Reprinted with permission from Synopsis, Inc.

in the physical environment can also perform placement-based scan chain reordering.

2.3.2 Power-Aware DfT

Adherence to power dissipation requirements is critical for design closure (e.g., for portable high-speed products). Static and dynamic power consumption and energy dissipation determine power hot spots on silicon. Power management and DfT must ensure that, for example, clock-gating circuitry would not reduce the testability of the design, while test points with scan logic would not increase power consumption.

Design testability is proportional to the percentage of scan registers it contains. For a register to be part of a scan chain, its clock and set/reset signals must be fully controllable from the chip boundary to ensure that data can be shifted through the chain. RTL clock gating reduces power consumption by selectively "shutting down" unused portions of the netlist (e.g., [37]; of clock-edge-driven designs). Such combinational or sequential gating logic in the clock paths introduces a risk that the registers violate test DRs. The clock gating logic might prevent the registers from being controllable in order to shift test data through them but can be modified to adhere to test requirements. A dedicated signal (*test_mode*) asserted while test patterns are applied to the design will override the clock control signal (*EN*) and allow normal operation of the scan chains when shifting or capturing test data. The drawback is that constraints on the test mode signal might affect testability of the clock gating logic and finite-state machine (FSM) since it will prevent observability of these blocks.

The control point (CP) can be placed between the latch and the register bank (i.e., behind) or between the FSM and the latch (i.e., in front). Inserting the CP behind can lead to performance degradation due to the logic between the latch and the register bank and the exposure of the clock signal to the danger of corruption due to glitches in the control signal value.

A popular technique for power management is controlling supply voltages. Typically, scan chain assembly is based on clock domains and may include placement information. For scan chains with multiple voltage domains, one concern for scan assembly is power dissipation. Voltage domain crossing is generally not a concern if the voltage of the source cell is greater than that of the drain cell. However, shifting data through scan chains, with the voltage domain of the source cell lower than the difference in voltages of the drain cell and the PMOS (P-channel MOSFETs) across the source and drain, would make the input PMOS transistors retain residual power that could lead to a short. To resolve this problem, level shifters with n-channel MOSFETs transistors are introduced at voltage domain boundaries. To minimize the area overhead due to shifter cells, multivoltage aware assembly orders the cells to minimize the chains that cross voltage domains.

Multivoltage aware scan chain assembly reduces area overhead due to fewer level shifters, power dissipation due to fewer voltage domain crossings, and wire length and routing congestion since cells in a voltage domain are ordered based on placement information.

DfT logic added to the design and scan registers assembled into scan chains could consist of simple structures like test points or complex structures for BIST, memory BIST, test controllers, test wrappers, and so on. Keeping logical and physical constraints in mind, logic must not increase power consumption. Low-power DfT includes multimode architectures, scan splitting, power-efficient scan cell designs (retention and gated scan-out flip-flops), power-saving logic on observe XOR (exclusive or) trees, and more. For test pattern sets applied to the physical device through ATE, test pattern sequences must minimize switching activity during test application. From a DfM standpoint, a key aspect of DfT is its impact on layout. TetraMAX can read the Synopsys physical database (Milkyway) [37] associated with physically clean design and display the corresponding physical data for the resulting fault candidates in its diagnosis output. The Verilog netlists and the physical database must match closely; instance names must match exactly.

In conclusion, critical issues in manufacturing testability related to the relentless growth in ASIC (application-specific integrated circuit) and SoC size and complexity, make DfT closure a mandate for the RTL to GDS (graphic database system) flow [38]. Along with function, timing, area, and power integrated design and test, EDA tools should migrate any potential test-induced iterations to the earliest stages of the design flow and comprehend a "test-friendly" RTL design [39].

2.4 Yield

High manufacturing yield is often considered both the key success criterion and key metrics of DfM. DfY is becoming the primary goal of product engineering [40]. There are two major causes of yield loss: random defects, often considered to be due to particles or other nonreproducible hot spots, and systematic defects due to the poor fidelity or incomplete process models when compared to the silicon data. Therefore, DfY and have yield to be considered in both roles: as a global deliverable of DfM with a relatively short and inexpensive feedback loop (Figure 2.3), which also predestines yield to be a success metric such as reliability and test ability with lower cause-effect ambiguity compared to less-tangible goals of DfM, but those standing higher in the hierarchy of rule of 10 (i.e., DfR and DfT). Accordingly, one can improve yield by a 1-man-day touchup to the layout and potentially confirm quantitative success in fab in a few days. This situation is much less likely (10–100× less) for DfR and DfT, which are not only not as quantitative but also more cross dependent on die architecture and product application. For this reason, we discuss yield aspects twice: in this chapter, as a DfM deliverable, and in Chapter 5, as a DfM metric. Yield is a dynamic deliverable, and to use it as a metric, one has to factor in its dependence on testing conditions and time:

$$Yield = \frac{N_G(Test,t)}{N_{Total}(throughput)} \tag{2.17}$$

Parameter	Description
N_G	Number of "good" (passing) dice according to testing criteria (Test) and time (t)
N_{Total}	Total number of dies depending on manufacturing throughput

Selection of (*Test*, *t*) depends now on the product setup. While there are several important categories of yield in IC manufacturing (Table 2.15), the ultimate goal of DfM is actually to maximize not just yield, but also yield-based product revenue:

$$R_{Revenue} = ASP(Test,t) \cdot N_G(Test,t) \tag{2.18}$$

Parameter	Explanation
$R_{Revenue}$	Total revenue of product
ASP	Average selling price depending on passing test bin
$Test$	Type of test
t	Time during product manufacturing period

While supported by yield, revenue is also a function of test conditions and time. DfY has an overruling goal which is designing for revenue, depending

TABLE 2.15

Categories of IC Yield

Yield Category	Abbreviation	Percentage of Passing...
Parametric	PY	Device parameters tested on scribe line structures
		IC parameters measured for sort yield
Functional	FY	IC parameters responding to test vectors
Sort	SY	Parameters at functional test at end of line
Class	CY	IC parameters depending on testing conditions
Systematic		Systematic dependence on manufacturing process window
Random		Depending on sudden parametric excursion or particle defect

on *ASP* which, in turn, is ensured by the testing conditions specific to the product function, with the extreme value point identified by

$$\frac{\partial R_{Revenue}}{\partial test} = \frac{\partial ASP}{\partial test} \bullet \frac{\partial N_G(Test, t)}{\partial test} = 0 \qquad (2.19)$$

Accordingly, DfY may have different meaning for the different manufacturing entities. For foundries, DfY rules have to ensure minimum pass criteria, to claim the highest yield possible, and to sell the maximum number of wafers. For captive fabs, which have to provide working ICs for customers with nonstandard yield testing for their specific markets, DfY is about designing in flexibility for the different products. Depending on the IC application, in high-reliability devices such as satellites, automotive, or aircraft industry products versus commodities such as disposable products, it could make more sense to run more wafers and sell a few selected dice at premium prices rather than try to build only a few wafers with a class yield close to 100%. Therefore, a trade off goal of DfY is to ensure no catastrophic defects at test. If CBC principles are observed, the many types of yield should be equivalent to one another [41]:

$$Y_{parametric} \qquad Y_{sort} \qquad Y_{class} \qquad (2.20)$$

with the minimal differences (yield suppression) due to the variability within the die, wafer, or lot.

Online yield for critical parameters lower than 100% indicates a misprocess and is not acceptable, so the wafers are scrapped. For PY to be an accurate predictor of SY, the mismatch between the DRs, device models, and the actual process performance should be eliminated. The mismatch occurs when process parameters, originally considered of secondary importance, start having an impact on the primary function of the device. For example, die planarity driven by pattern density distribution has an impact on capacitive coupling, causing location-dependent frequency response of RF devices. PY below 100% is a result of a gap between DRs and the required process capability [42]. A violation of DRs signifies a design defect that has to be fixed. But, a design marginally passing the DRC check can still show failures to the yield bins due to a number of failure modes. Therefore, missing or too-forgiving rules must be identified. For example, a DR regarding the minimum width of polysilicon line, if set up below the actual process capability, could result in multiple failure modes of the product, such as MOSFET leakage or breakdown, high poly contact resistance, or low ESD tolerance. A correlation matrix can be set up between rule marginalities and yield loss mechanisms:

$$f(V\_DR_i(x,y)) = FM_{i1} \quad \cdots \quad FM_{in} \qquad (2.21)$$

Parameter	Explanation
$V\_DR_i(x, y)$	Event of DR marginality (violation) for DR: rule number = i, at die layout coordinates = x, y
$FM_{i1...in}$	Failure modes due to V_DR total n failure modes possible
$\otimes$	Logical "or" operand

Model parameters linking these violations to yield loss should then be extracted. A more difficult situation arises when the rule is not present in the deck and has to be derived based on the circumstantial evidence from yield analysis. The criteria for such a rule would at first be defined by the layout context, followed by understanding of the root cause and applying a permanent correction.

But, DfY must also focus on resolving failure modes beyond the scope of checkable DRs [43]. For example, minimum line CD (critical dimension) can be checked on the layout but not ensured in all locations on all wafers due to inadequate pattern transfer quality. OPC (optical proximity correction) models, simulated edge placement errors, or CD measurements are not always able to find all the drawn-to-silicon CD discrepancies (e.g., due to 3D effects such as reflective notching at active-isolation interface). As a corrective action for such situations, either the minimum poly CD DR must be tied to its context, such as the proximity to the isolation edge:

- minimum poly CD within more than 100 nm of active edge = 100% of nominal rule

- minimum poly CD within 100 nm or less of active edge = 120% of nominal rule

or the proximity effect has to be corrected by CAD (computer-aided design) postprocessing using tools capable of simulating the process.

Test cases for DR verification (violator cells) are becoming more complex with IC architecture development. The number of tests a product must pass should correspond to DR implementation via a correlation matrix linking the process and the electrical deliverables of the product with their checkable layout.

If CBC and rule of 10 were always followed, device functionality would be confirmed first at the interconnect level (i.e., at metal 1). The first PY read point should be sufficient (e.g., for the known good die [KGD] methodology) [44]. DfM for PY is impacted by not only 2D, but also 3D effects in LfD (litho-friendly design), such as photoresist shadowing, and by parasitic RLC effects, which can be modeled across frequency ranges at the die (or product) level. Yield reduction due to particles or other random defects is mitigated by adding redundant blocks (e.g, for static random access memories [SRAMs]) to the extent permitted by RoI expectations.

Low-risk DfY redundancy schemes, not requiring design intervention or repair, are called for to help eliminate random defects. For example, doubling contacts and vias by a CAD algorithm is a popular DfM/DfY best practice. But, its application would lead to nondeterministic design performance. A similar layout, depending on its environment, may or may not double the vias, creating sensitivity issues:

1. Electrical: Would the extra vias cause problems (e.g., with device matching due to the different parasitics) (RC)?
2. Physical: Would the via enclosures required by the overlying and underlying metals increase the critical area or sensitivity to random defects on those layers?

A safer approach, and not necessarily a more expensive one, compared to redundancy, is a high-quality manufacturing line. Standard yield models reflecting the impact of random defects should help minimize the need for redundancy.

Systematic yield loss can also be modeled with accuracy depending on model quality (i.e., extraction) for optical, physical, or electrical effects, such as

- Pattern reproduction on wafer: Does it correspond with desired accuracy to the simulated shapes?
- MOSFET current-voltage (I-V) characteristics: Does the sub-V_T leakage model correspond to product requirements and in-line data?
- Is the dielectric deposited over topography "perfectly" flat after CMP?

In the case of a mismatch between the model, the inline data, and the product parameters having an impact on product applications and revenues, the models have to be reconciled with reality

- by improving our understanding of the physical phenomena, capturing higher-order effects in equations and computing tools, and locking the manufacturing parameters, or
- by modifying the manufacturing process to match the models and ensure process stability.

For example, matching the on-wafer pattern to its model may require

- developing advanced litho simulations and OPC techniques, or
- switching exposure wavelength from 193 nm to EUV (extreme ultraviolet) or direct e-beam writing where pattern transfer would not be subject to distortions that have an impact on model fidelity.

Model development would eventually meet limitations fundamental enough to reduce the RoI of further efforts to reconcile with previous process implementation below the break-even point. One can try to use the 193 nm exposure wavelength indefinitely down the CD shrink path proposing ever-more-complex simulation engines and assist features. But, at one point, adapting new technology is the only viable solution, even though the industry is attempting to push out as far as possible the onset of new process paradigms due to the associated cost.

2.4.1 Parametric Yield

Parametric yield (PY) can be defined as [43]

$$PY = \frac{Number\ of\ passing\ critical\ device\ parameters}{Total\ number\ of\ critical\ device\ parameters} \quad (2.22)$$

The goal of a PY check is to confirm that DR criteria assumptions are correct. Therefore, PY less than 100% is usually not acceptable, and deviations should be identified as early in design and manufacturing flows as possible. PY can also be tracked to reflect IC response to controlled process variations. Digital circuits are typically optimized only for speed and power, while analog circuits have to comply with many other performance metrics, and their process variations require higher complexity of yield modeling. Simulations account only for a fixed process point, but they fall short of optimizing PY response to process variations and worst-case environmental factors (e.g., V_{dd} and temperature) for each point in the process. For analog/mixed-signal or memory design, the worst-case condition may be different for each performance parameter, such as gain, bandwidth, and jitter, and design cannot be simultaneously optimized to ensure high yield for all of them, which is always lower than the lowest partial yield for each parameter separately.

Designers verify the operation of their circuits at the center (nominal) point of the process and at the typical four worst-case "corner" points corresponding to the typical fail control range. But, without sweeping across the full range of process and operational parameters, revenue loss attributable to PY can be on the order of 20%. Accordingly, a product may need 10,000 extra wafers over the 3-year lifetime for a total cost of $25 million, at $2,500 per wafer, because of its poor yield.

Efficient DfY requires an automated way to maximize the distance between the nominal process point and the point at which the product begins to fail, referred to as the "worst-case distance." By knowing the worst-case distance for each product and process parameter related to it, the devices in the circuit can be optimized to make them tolerant of the process sensitivity.

After the improved design is placed in the fab, parametric data collection systems need to provide feedback about parametric response to design adjustments. The storage and transfer of these data among the different IP providers (design houses) and different manufacturers (foundries) related to DfM optimization has to keep in mind communication, cross-fab compatibility protection, and cost reduction. These procedures are not standardized at this point.

2.4.2 Functional, Sort, and Class Yield

Ensuring high functional sort and class yields is only possible through process architecture based on CBC aligned with accurate modeling and reliable components in addition to DfR, LFL (litho-friendly layout), DfT, and so on. The design expectation is to have a design environment that would help to deal proactively with those issues [45].

Such a set up should be supported by yield models reflecting into the DR space. Converting single-value DRs into yield models based on the distributions of pattern properties extracted from layout features would not only help meeting all standard DRs (e.g., printability, isolation, pattern density) but also ensure yield close to 100%, assuming that the extrinsic process defect density $D_0 = 0$. Process marginalities, not fully reflected in the basic DR set, require statistical analysis to better respond with layout options. For example, yield improvement due to wire spreading, via doubling, pattern density equalization, or mask complexity reduction, can often be attained by locally expanding layout geometries beyond their required minimum CDs defined by the DRs, to the larger dimensions, reflected by the recommended rules (RRs). For DfM-based improvement yield, models can make it a turnkey solution by correlating layout features with yield loss due to critical process constraints, which for the current generation of the ICs are lithography, particle, and planarity. Layout enhancements to mitigate these constraints were found to work better when performed in a local environment, case by case, rather than by automated routines at global (chip) level.

Layout modifications for DfY should rely on the following best practices to preserve design intent:

- preserve critical nets and use collapsible netlists
- use as few dummy features as possible for CMP planarization
- preserve critical features and provide metrology guidelines for critical area analysis (CAA)
- keep pattern density gradient as low as possible (no "grand canyons" in die topography)

Model-based design would tighten parametric distributions for products with multiple types of active and passive elements (transistors, resistors, capacitors, inductors). Evolution of product families is enabled by process enhancements (line-width scaling), which improve DR margins, or by the new product requirements, which increase DfM complexity, or both. The resulting explosive growth of DR count (by about 20× over the last decade) (Table 2.16) should be mitigated by a model-based approach [46]. Here, the DR and RR rules integrate into process-based DfM models to ensure, for example, device matching, not easily implementable by either the DR or RR approach alone. In the process of model definition for DfY, one would first derive equations linking die geometries with yield, followed

TABLE 2.16
Evolution of Technology Nodes, Process and Device Options, and Design Rule Numbers

Example Product Introduced at Process Node	Memories	Complex Logic	CMOS Analog	RF	SoC	Issues/Limitations
Lithography	0.25 or more	0.25–0.18	0.18–0.13	0.18–0.09	0.13–0.065	LLY
Conducting layers	3	5	7	9	12	
Planarization and isolation technologies	Deposition/etch and LOCOS	Deposition/etch, CMP, LOCOS, STI	CMP, STI	Advanced CMP, STI	Advanced CMP, DTI	CMPLY
Fab class	>100	100>	>10	10>	1	PLY
Devices	Small MOSFETs	Small MOSFETs requiring OPC	Precision small and large MOSFETs requiring OPC, resistors	Precision MOSFETs, resistors, capacitors, inductors	Precision active and passive elements, routing	Wide variety of layouts difficult to optimize
Number of design rules	<100 (<5/layer)	<200 (<10/layer)	<300 (<15/layer)	<350 (<15/layer + global matching rules)	<500 (<20/layer + custom device and matching rules)	Large number of rules/values

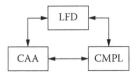

FIGURE 2.25
Cross-dependence among CAA, LFD, and CMP optimization. (From A. Balasinski, F. Pikus, and J. Bielawski. 2008. Advanced Semiconductor Manufacturing Conference. *IEEE.* With permission.)

by calibrating them with process parameters related to three key process defect components:

- lithography defects, limiting lithography yield (LLY), mitigated by LFD
- particles, limiting particle yield (PLY), mitigated by CAA [47]
- planarity defects, due to the nonuniform pattern density distribution over the die area, limiting CMP yield (CMPY), mitigated by pattern density limits for planarization (CMPL)

Because layout optimization requires different approaches to maximize LLY, PLY, and CMPY, a question is how the defect-specific optimization would impact other types of defects in the yield improvement scheme. The three key process defects—particles, poor pattern fidelity, and nonplanarity—show some correlation (Figure 2.25) with a strong correlation between DfM approaches for LFD and CAA [48], allowing proposals of empirical formulas to express such correlation.

Layout that improves yield for one type of process defect may degrade it for another.

For example, improving PLY by doubling vias increases critical areas of via enclosures and may compromise CMPY of metals. Bending and spreading metal lines to reduce particle sensitivity increases the lithographic yield sensitivity. Redundant vias reduce via failures but increase critical metal areas and PLY sensitivity. High LLY requires consistent layout pitch in one direction, while layout optimized for PLY and CMPLY should be isotropic to even out the impact of medium- or long-range interactions. Both LLY and PLY guidelines recommend using minimum CD only where it is important for speed or area, and larger CDs where the area does not need to be spared to reduce not only process variability but also the number of dummy structures to fill the empty space. As the best compromise, layout modifications should be performed on a small scale such that the highest yield can be achieved by selecting different types of layout enhancements in different locations of the die [46].

The DfM defect metric for LLY, PLY, and CMPLY related to wire spreading, via doubling, and extra enclosures (Figure 2.26) shows higher DfM values,

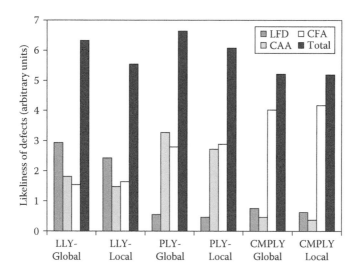

FIGURE 2.26
DfM metric for three types of layout optimization. Likeliness of defects on a 7-point scale (7 = very likely; 1 = very unlikely). (From A. Balasinski, F. Pikus, and J. Bielawski. 2008. *Advanced Semiconductor Manufacturing Conference. IEEE.* With permission.)

corresponding to higher probability of defect improvements where local optimization overrules global optimization for LLY and PLY. On the other hand, global and local optimization are equally effective for CMPLY due to the long CMP interaction range. Because global optimization is not sensitive to the local defects, local optimization is much more effective in improving DfM quality, being also more time consuming, due to case-by-case reviews and fixes at a low hierarchical design level. Accordingly, yield models need to be extracted to include local defects. Neither simple increase of layout footprint nor geometric rule checks are sufficient DfM solutions anymore to ensure high LLY, PLY, and CMPLY yields. One should use model equations for LFD and CAA optimization on a local scale and global optimization for CMPLY.

For the LFD and CAA, yield is correlated to the CD pitch (Figure 2.27): With CDs above the minimum pitch for technology, the lithography yield -> 100%. With defect density at the extrinsic process value of D_0 (i.e., related to particles), a shift in DR from minimum (x_{min}, y_{min}) to recommended values $(x_{min} + dx, y_{min} + dy)$ (RR) may increase circuit area at a quadratic rate $(x_{min} + dx)*(y_{min} + dy)$ (Figure 2.27b). As a result, particle-related yield loss based, for example, on a Poisson yield model begins to dominate over other yield loss mechanisms. The resulting profitability can be expressed as a function of die footprint such that it would follow the product of PLY, LLY, and the die footprint itself, and reaching a maximum at the optimal RR value (Figure 2.27d).

Pursuing trade-offs in multiparameter layout DfM optimization is a difficult task. For example, CMPLY is a function of both absolute value and

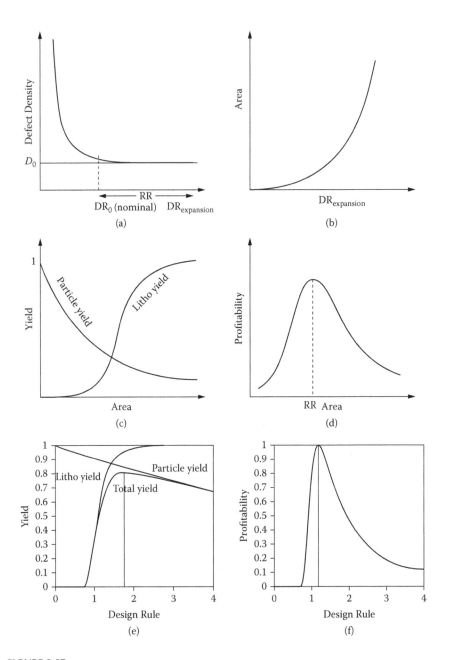

FIGURE 2.27
Relationships among design rules, defect density, die area, yield, and profitability. Correlations:
(a) defect density design rule; (b) area-design rule; (c) yield vs. area (d) profitability vs. area; (e)
yield vs. design rule; (f) and profitability vs. design rule. (From A. Balasinski, F. Pikus, and J.
Bielawski. 2008. Advanced Semiconductor Manufacturing Conference. *IEEE*. With permission.)

gradients of pattern density (PD). It decreases sharply when ΔPD exceeds a certain value. For low PD, the polishing time is typically shorter, and the CMP process would be less likely to result in overpolishing of the ILD so the potential yield drop is smaller for PD on the order of 30% than for PD on the order of 60%.

A common method to improve CMPLY is by adding fill pattern. This has to be modeled independently from LFD and CAA. PD rules may first require adjusting drawn geometries because adding fill pattern to any preexisting layout may not always be effective.

CMP optimization may be based on three different types of fill pattern such as (Figure 2.28)

- "intelligent" fill, which adjusts itself to pattern density of the neighboring areas, ensuring low impact to sensitive analog/RF devices
- manual fill, which can be fully extracted if required by the RF circuits
- geometric fill of fixed density over the die area

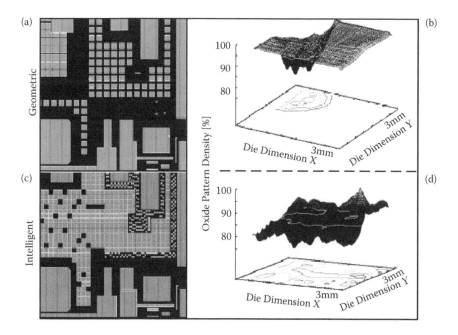

FIGURE 2.28
Fill pattern impact on pattern density (a) sparse geometric waffling; (b) 3-D image of pattern density distribution; (c) dense "intelligent" waffling; and (d) 3-D image of pattern density distribution. (From A. Balasinski, F. Pikus, and J. Bielawski. 2008. Advanced Semiconductor Manufacturing Conference. *IEEE*. With permission.)

One can propose generic fitting formula integrating the expected correlations [45] depending on the value and tolerance of DR,

$$Y_i(x) = \left[1 - \exp\left(-\frac{x - (DR_i - 3\sigma_i - tol_i)}{DR_i - 3\sigma_i - tol_i}\right)^n\right] \cdot \exp(-1/x)^n \qquad (2.23)$$

Parameter	Explanation
x	Value of a DR (independent variable)
DR_i	Proposed value of DR DR_0 for the ith masking process step
tol_i	DR tolerance for the ith masking process step
$3\sigma_i$	Variation of the DR for the ith masking process step
n	Sensitivity factor of yield to the change of DR value

Coefficients in this equation should be extracted from the layout-dependent process variations of product yield. The optimal value is identified as the peak in Figure 2.27. In a simpler approach, yield as a nonlinear function of DR value (x) can be modeled (e.g., by a sigmoid function of design rule only):

$$y(x_{DR}) = \frac{1}{1 + \exp(-x_{DR})} \qquad (2.24)$$

which increases from $y \approx 0$ for low x_{DR}, through a rapid growth when X_{DR} gets close to the minimum R assumed by technology, and saturates at 100% for larger x values, from which one can choose the optimal value of RR. The shape of the function can then be controlled by process capability and sensitivity factors:

$$Y(x) = 1 - \left(\frac{1}{1 + k \exp(-x_{DR} + PC)}\right)^n \qquad (2.25)$$

Parameter	Name	Comments
PC	Process capability	Related to the minimum design value of rule DR
k	Process sensitivity factor at minimum x value	Arbitrary units
n	Layout sensitivity factor across the range	Arbitrary units

In summary, DfM yield improvement combined with product development requires use of RoI-based trade-offs for die footprint optimization and model-based physical verifications.

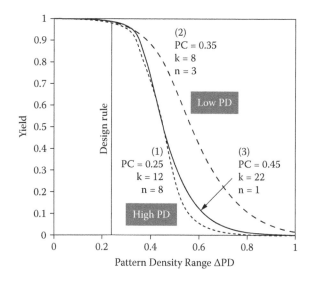

FIGURE 2.29
CMP yield as a function of pattern density range ΔPD for high- and low-absolute PD values. Curves (1), (2), and (3) correspond to the different sets of sensitivity and capability factors. (From A. Balasinski, F. Pikus, and J. Bielawski. 2008. Advanced Semiconductor Manufacturing Conference. *IEEE*. With permission.)

As an example, Equation 2.25 can be used to predict yield response to pattern density variations based on the following calibration coefficients (Figure 2.29):

- Process capability *PC* (i.e., PD range where planarization yield is ~100%) across the reticle ΔPD field, requires pattern density range not to exceed 25%. That means, DR = 0.25.

- High value of sensitivity factor *k* would correspond to an abrupt yield drop when ΔPD > DR. Concurrent changes in *PC* and *k* calibrate the yield model with fab data (e.g., for *PC* of 0.45, the *k* factor would increase from 12 to 22) to arrive at a similar yield curve. In other words, the further the DR value is pushed, the more rapid is the yield drop.

- *n* can be assigned to other layout parameters not captured in pattern density such as waffle size, shape, and space to keep out regions, etc., in DRs.

2.5 Summary

In this chapter, we discussed hierarchical implementation of DfM, from the most to the least-critical issues from the customers' standpoint, that is, from reliability, to functionality testing, to yield. The solutions to these issues

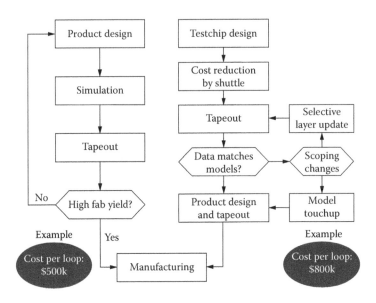

FIGURE 2.30
Example of test chip RoI calculation, based on the number of design operations.

are usually designed into the product in that order, although the execution sequence as well as the importance to the product line are often reversed: first high yield, then reasonable testability, then "good" reliability verified mostly on legacy products and limited-size experiments. Note that failure mechanisms related to reliability, methods to ensure adequate test coverage, and the origins of yield fallout depend on the IC technical solutions not likely to change in the future. The products would always need to be reliable from the mechanical (stress) and electrical (stress and discharges) standpoints, testable *in situ*, and show satisfactory yield response to several key process parameters, consistent with mathematical models.

A popular cost—and risk—reduction methodology is to run test chips before committing to the actual product. Adding a test chip into product flow would add a cycle of learning, potentially more expensive than that of a regular product (Figure 2.30). However, that cost, if distributed among several products, would be readily amortized typically if the number of product COLs (cycles of learning) is greater than one.

References

1. Anderson, D. M. *Design for Manufacturability and Concurrent Engineering.* Cambria, CA: CIM Press, 2008.
2. Moore, G. Cramming more components onto integrated circuits. *Electronics,* 38(8), 114–117, 1965.

3. MIL-STD-785, IEEE 1332.
4. http://en.wikipedia.org/wiki/Electromigration: Electromigration, July 2010.
5. Black, J. R. Electromigration—a brief survey and some recent results. *IEEE Transactions on Electron Devices*, 16(4), 338–347, 1969.
6. Migita, S., Morita, Y., Mizubayashi, W., and Ota, H. Preparation of epitaxial HfO$_2$ film (EOT = 0.5 nm) on Si substrate using atomic-layer deposition of amorphous film and rapid thermal crystallization (RTC) in an abrupt temperature gradient. *Proceedings IEDM*, 269–272, December 2010.
7. Teng, C. A. Dicing advanced materials for microelectronics. *Proceedings, International Symposium Advanced Packaging Materials, Processes, Properties and Interfaces*, 149–152, 2005, March 16–18.
8. Eriguchi, K., and Niwa, M. Correlation between lifetime, temperature, and electrical stress for gate oxide lifetime testing. IEEE Electron Device Letters, 18(12), 577–579, 1997.
9. Sahhaf S., Degraeve, R., Roussel, P.J., Kaczer, B., Kauerauf, T., and Groeseneken, G., A new TDDB reliability prediction methodology accounting for multiple SBD and wear out. *IEEE Transactions on Electron Devices*, 56, 1424–1432, 2009.
10. Prendergast, J., Suehle, J., Chaparala, P., Murphy E., and Stephenson M. TDDB characterization of thin SiO$_2$ films with bimodal failure populations. *IEEE Reliability Physics Symposium*, 124, 4–6 April 1995.
11. *Failure Mechanisms and Models for Semiconductor Devices*. JEDEC Publication JEP122E. JEDEC Solid-State Technology Association, March 2009, Solid State Technology Association Arlington, VA.
12. Hu, C., and Lu, Q. A unified gate oxide reliability model. *IEEE Reliability Physics Symposium Proceedings*, 47–51, March 22–25, 1999.
13. Franco, J., Kaczer, B., Eneman, G., Mitard, J., Stesmans, A., Afanas'ev, V., Kauerauf, T., Roussel, J., Toledano-Luque, M., Cho, M., Degraeve, R., Grasser, T., Ragnarsson, L.-Å. L., Witters, L., Tseng, J., Takeoka, S., Wang, W.-E., Hoffmann, T. Y., and Groeseneken, G. 6Å EOT Si0.45Ge0.55 pMOSFET with optimized reliability (VDD = 1V): meeting the NBTI lifetime target at ultra-thin EOT. *Proceedings IEDM*, 70–73, December 2010.
14. Rosenbaum, E., Rofan, R., and Hu, C. Effect of hot-carrier injection on n- and pMOSFET gate oxide integrity. *IEEE Electron Device Letters*, 12(11), 599–601, 1991.
15. Hu, C. Lucky-electron model of channel hot electron emission. *Proceedings IEDM*, 22–25, December 1979.
16. Groeseneken, G., Degraeve, R., Nigam, T., et al. Hot carrier degradation and time-dependent dielectric breakdown in oxides. *Microelectronic Engineering*, 49, 27–40, 1999.
17. Li, X., Huang, B., Qin, J., Zhang, X., Talmor, M., Gur, Z., and Bernstein, J. B. Deep submicron CMOS integrated circuit reliability simulation with SPICE. Proceedings of the Sixth International Symposium on Quality Electronic Design, 21–23 March 2005.
18. Krishnan, A. T., Cano, F., Chancellor, C., Reddy, V., Qi, Z., Jain, P., Carulli, Masin, J., Zuhoski, S., Krishnan, S., and Ondrusek, J. Product drift from NBTI: guardbanding, circuit and statistical effects. *Proceedings IEDM*, 78–81, December 2010.
19. Schroder, D., and Babcock, J. A. Negative bias temperature instability: road to cross in deep submicron silicon semiconductor manufacturing. *Journal of Applied Physics*, 94(1), 1–18, 2003.

20. LaRosa, G. NBTI challenges in PMOSFETs of advanced CMOS technologies. IEEE International Reliability Physics Symposium, Tutorial Notes, 2003.

21. Methods for Calculating Failure Rates in Units of FITs, JESD85. JEDEC Publication. JEDEC Solid State Technology Association, Arlington, VA 2001.

22. Srinivasan, J., Adce, S. V., Bose, P., et al. *The Case for Microarchitectural Awareness of Lifetime Reliability.* University of Illinois at Urbana Champaign, Computer Science UIUC CS Technical Report, 2003.

23. Wu, H., Archer, V., Merchant, S., Cargo, J., Chesire, D., Antol, J., Mengel, R., Osenbach, J., Horvat, S., Peridier, C., and White, M. Advanced failure analysis of circuit-under-pad (CUP) structures in Cu/FSG and Cu/low K technologies. IEEE 43rd International Reliability Physics Symposium, April 17, 28, pp. 286–293 2005.

24. Fukuda, Y., Ito, S., Ito M., et al. *Technical Review 185, Vol. 68, Special Edition on 21st Century Solutions, SOI-CMOS Device Technology,* 2001.

25. Haseloff, E. *Application Report SLYA014A—Latch-Up, ESD, and Other Phenomena,* 2000. Texas Instruments, Dallas, TX, USA.

26. Vinson J. E., and Liou, J. J. Electrostatic discharge in semiconductor devices: an overview. *Proceedings of the IEEE,* 86(2), 399–420, 1998.

27. Department of Defense Test Method Standard, 1996. MIL-STD-883E.

28. Logical Link Control Standard IEC 802-2.

29. ITRS Technology Roadmap.

30. Nag, P. K., and Maly, W. Simulation of yield/cost learning curves with Y4, *IEEE Transactions on Semiconductor Manufacturing,* 10(2), 256–266, 1997.

31. De Colle, A. *Test Process Phases and Guidelines.* Synopsys, 2006. Mountain View, CA.

32. Hayat, F., Williams, T. W., Kapur, R., and Hsu, D. DFT closure. *Proceedings of the Ninth Asian Test Symposium,* 4–6, December 2000.

33. Hirech, M., and Ramnath, S. Moving from one-pass scan synthesis to one-pass DFT synthesis. IEEE European Test Workshop, 3B.1, 2001.

34. Rajski, J., and Tyszer, J. *Arithmetic Built-In Self-Test for Embedded Systems.* Englewood Cliffs, NJ: Prentice Hall, 1998.

35. *DFT Compiler User Guide: Adaptive Scan (XG Mode),* Vol. 2, Release 2006.06. Synopsys, Mountain View, CA 2006.

36. *IC Compiler User Guide,* Release 2006.06. Synopsys, 2006.

37. *Library Compiler Reference Manual: Physical Libraries,* Release 06.2006 Synopsys, 2006.

38. Murphy, C. F., Abadr, M. S., and Sandborn, P. A. Economic analysis of test process flows for multichip modules using known good die. *Journal of Electronic Testing: Theory and Applications,* 10, 151–166, 1997.

39. Design for test. 2011. http://en.wikipedia.org/wiki/Design_For_Test.

40. Chiang, C., and Kawa, J. Design for manufacturability and yield for nano-scale CMOS integrated circuits and systems, design for yield, Springer 2007.

41. Reda, S., Smith, G., and Smith, L. Maximizing the functional yield of wafer-to-wafer 3-D integration. *IEEE Transactions on Very Large Scale Integration (VLSI) Systems,* 17(9), 1357–1362, 2009.

42. Ferri, C., Reda, S., and Bahar, R. I. Parametric yield management for 3D ICs: models and strategies for improvement. *ACM Journal on Emerging Technologies in Computing Systems,* 12, 1–20, 2008.

43. May, G. S., and Spanos, C. J. *Fundamentals of Semiconductor Manufacturing and Process Control.* New York: Wiley, 2006.
44. Yip, W. K., Law, K. G., and Lee, W. J. Forecasting final/class yield based on fabrication process E-test and sort data automation science and engineering. *IEEE Automation Science and Engineering Conference,* 478–483, 2007.
45. Balasinski, A., Pikus, F., and Bielawski, J. Yield optimization with model based DFM. *IEEE/SEMI Advanced Semiconductor Manufacturing Conference, ASMC 2008,* 216–220, May 2008.
46. Pikus, F. G., Andres, J., and Torres, A. Non-uniform yield optimization for integrated circuit layout. *Proceedings SPIE,* 6730, 67300Y-I-Y-12, 2007.

3

IC DfM for Devices and Products

3.1 Introduction

Integrated circuit design for manufacturability (IC DfM) methods and tools can be applied at different levels of design hierarchy, from individual devices, through standard cells, IP (intellectual property) blocks of various functions, up to the chip level. DfM principles for individual devices would focus on ensuring their alignment with models, lowering the parasitics, and improving reliability. DfM for generic metal oxide semiconductor field effect transistor (MOSFET) parametric cells (pCells) would deal with reducing the length-of-diffusion effect [1], optimizing the area-to-perimeter ratio, and accounting for electromigration (EM). At the IP block level, DfM methodology should be concerned about pattern reproduction across the process window in addition to verifying and integrating solutions from the lower levels.

From the design architecture standpoint, cell DfM needs to improve its transferability to the next technology generation or another family of devices. At the chip level, DfM needs to consider multidimensional effects with the degree of complexity increasing from one-dimensional (1D) scalability through global two-dimensional (2D) pattern evaluation and hot spot reduction with litho-friendly design (LfD), chemical-mechanical polishing (CMP), and critical area analysis (CAA) methods; to 3D, for which vertical product architecture and functionality are decided; and finally into 4D, for which the design longevity (reliability, but also versatility) decide on its ultimate success in the marketplace. DfM methods (correct by construction [CBC], design for reliability [DfR], design for yield [DfY], etc.) and tools (CAA, LfD, OPC, etc.) can be introduced at all these different stages of design flow [2]. Emerging DfM at the product level is oriented at the packaging optimization for market applications.

Typically, design of a new part starts from improving the legacy design element for entire products. Optimizing them using new tool sets or adjusting for the different applications often requires DfM intervention. Critical questions for such intervention include the preferred DfM approach: to correct or to stabilize, which translates into aggressive or conservative DfM solutions. The first type introduces an additional degree of freedom (e.g., layout correction features, which may destabilize the process window but salvage the existing design); the second type would stabilize the process window at the expense of design parametrics.

In this chapter, we discuss

- DfM for standard cells: memory and logic
- DfM at product definition: system in package (SiP) versus system on chip (SoC)
- Variability reduction at 1D, 2D, 3D, 4D:
  - 1D: extracting process parameters from a fixed-pitch pattern
  - 2D: following design intent by ensuring printability
  - 3D: ensuring planarity of layers or optimizing vertical product architecture
  - 4D: understanding time dependence and reliability

Product architecture decides about the preferred path of DfM implementation: top-down or bottom-up (i.e., from device to circuit to product or from product to circuit to device). While in the end, it all matters equally—DfM for devices (active, passive, interconnects) and at a product level—it is the proper hierarchy of DfM implementation that would make it easier to identify DfR, DfY, or litho-friendly layout (LfL) hot spots. For products integrated from a small number of standard cells, a bottom-up approach helps achieve higher layout density, while for products hosting mainly random logic, top-down DfM may help save development time.

3.2 DfM for MOSFETs and Standard Cells

One simple starting point for DfM implementation is to ensure controllable operation of active devices (MOSFET) in digital circuits as ON/OFF switches. While they need to function predictably and reliably within their predefined input/output (I/O) parameter set, that set becomes increasingly more aggressive from one technology and product generation to the next. Luckily, the simplicity of MOSFET layout—a conductive path (e.g., active area) and a valve (gate) running across it—makes the transistor footprint easily shrinkable in planar technology (2D), while the device parameters across a wide range of dimensions (from nm to mm) respond well to the product needs (Figure 3.1). MOSFET manufacturability challenges can readily be linked to the parameters of its basic model, reflected in the formula for drive current in saturation [3]:

$$I_D = \mu \frac{W \cdot K}{L \cdot t_{ox}} (V_{GS} - V_T) \cdot \frac{V_{DS}^2}{2} \qquad (3.1)$$

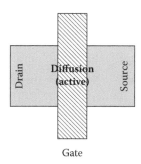

FIGURE 3.1
Basic MOSFET layout (2D).

TABLE 3.1

Parameters with an Impact on MOSFET Drive Current

Parameter	Name	Factors Impacting Model Accuracy
I_D	Drive current	All parameters of the equation
μ	Carrier mobility	Mechanical stress and temperature, layout context, and device operating conditions
W	Channel width	Layout proximity impact on the printing process
L	Channel length	Roughness of gate edges due to gate material and line definition
		Layout proximity impact on the printing process
K	Dielectric constant of gate oxide	Spatial distribution of oxygen, nitrogen, dopant atoms depending on the thermal process
t_{ox}	Gate oxide thickness	Atomic-scale thickness variations
VGS, VDS	Gate-source voltage	DC and AC IR drop over connecting lines: PCB, interblock routing, local interconnects
VT	Threshold voltage	Channel geometry and doping distribution

Despite the apparent simplicity of the model described by Equation 3.1, none of its parameters (Table 3.1), although assumed to be constant for the given process or directly controlled from the inputs of the IC, is always known with sufficient accuracy to ensure fully predictable functionality of all MOSFETs in an IC product over its specified lifetime. One should notice that this MOSFET model does not directly include second-order, nonuniform, 3D, tunneling, and time-dependent (reliability) effects such as

- shallow-trench isolation (STI) stress (impacting mobility) [4]
- gate-induced drain leakage (GIDL; impacting OFF-state leakage current) [5]

- SILC (stress-induced leakage current; impacting reliability and low standby current) [6]
- poly depletion (impacting gate turn on and off) [7]
- shallow source/drain junctions (impacting punch-through) [8]
- negative bias temperature instability (NBTI; impacting reliability) [9]

which makes it virtually impossible to predict MOSFET behavior with accuracy mandated by the many complex products, as processed (in 3D) or over time and voltage domains (in 4D). As a DfM response, to capture second-order effects, multiple parameters would have to be extracted from the measurements, fed into distributed static models, and converted into experimental formulas. The dynamic MOSFET behavior, impacted by RLC components, would be even more complicated. Therefore, predicting device behavior is even more difficult for analog/RF (radio-frequency) circuits, where corner modeling no longer suffices, but discrete models are required. The problem of model accuracy is exacerbated for subthreshold MOSFET operation described by the following equation [3]:

$$I_D = I_0 \frac{W \cdot K}{L \cdot t_{ox}} \exp(V_{GS} - V_T)/VC \cdot (1 - \exp(qVDS/kT)) \exp\left(\frac{V_{es} - V_T}{V_C}\right) \cdot \left[1 - \exp\left(\frac{qV_{DS}}{kT}\right)\right]$$

$$(3.2)$$

which illustrates that MOSFET models are facing all the uncertainty challenges enhanced by the exponential factor. Therefore, despite many advantages, operating MOSFETs in the subthreshold range is not a good manufacturability proposition [10]. Low-cost 3D process architectures such as FINFETs are a better option to extend the complementary metal oxide semiconductor (CMOS) road map [11].

From the standpoint of device performance. DfM can enhance one of two options reflected in Equation 3.1. Aggressive DfM would maximize output current by enabling low L and parasitics for RF applications, whereas conservative DfM would ensure device reproducibility across wafer and lot, even if at the expense of die footprint and parametric data.

But, to decide on the right approach, the quality of device layout with an impact on the quality of models needs to be assessed first by simulations of silicon image across the process window of pattern reproduction technology (e.g., lithography).

It is not possible to build models without highly accurate reproduction of drawn geometries on the wafer, reflecting design intent. As the ITRS (international technology roadmap for semiconductors) road map [12] offers numerous solutions to ensure matching of drawn and silicon patterns, the design databases need to be verified by pattern transfer simulation to match

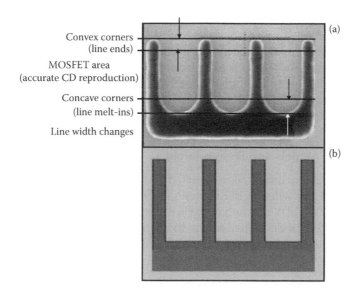

Convex corners
(line ends)

MOSFET area
(accurate CD reproduction)

Concave corners
(line melt-ins)

Line width changes

FIGURE 3.2
SEM image (a) and drawn layout (b) of a multi-finger MOSFET showing line width changes at minimum spacing of the gate to areas where CDs must correspond accurately to drawn values the inside (concave) corners and minimum spacing of the gate to the line of extended width (convex corners).

criteria for the desired device geometries to ensure both performance and manufacturing tolerance of the circuits.

For optical lithography, pattern fidelity of silicon images representing drawn polygons (Figure 3.2) depends on the feature size and shape and on the exposure parameters: refractive index (n), numerical aperture (NA), and wavelength λ.

When device dimensions continue to scale along the milestone nodes (90, 65, 45, 28, 22, 15 nm), all other process modules and techniques also need to respond to this trend to keep a satisfactory quality of device models. Pattern transfer techniques were at one point qualified by what it took to achieve best "naked eye" agreement between the drawn layout and the silicon image. To verify advanced lithography techniques, geometry distortions are now expressed in terms of distributions of edge placement errors (EPEs) over the die area [13]. EPE calibration to a simulated silicon pattern is based on critical dimension (CD) measurements at locations where drawn CDs must accurately correspond to the final ones over the exposure/defocus (ED) window. In those locations, the EPEs must be linked to device parameters and stay within specified limits.

But, the spot checking of critical CDs to the simulations may not suffice when devices of different dimensions and functions have to be accurately reproduced on silicon. Focusing on the smallest CDs is no longer sufficient as IC failures may be caused by devices with longer channels and operating at higher voltages. This, combined with the increasing number of approaches to pattern transfer enhancements (optical proximity correction [OPC], inverse lithography technology [ILT], source mask optimization [SMO]) [14,15], as well as manufacturability requirements across manufacturing sites, calls for more comprehensive methods to capture design intent and make it available for the manufacturing line, to support the infrastructure for pattern quality improvement developed among design, computer-aided design (CAD), mask, and manufacturing organizations.

3.2.1 Device Models and Design Intent

Comparing design intent represented by drawn layout to silicon pattern became important when light diffraction caused a noticeable deviation from the expected "WYSIWYG" (what you see is what you get) image (i.e., at about the 1-μm process node). This issue is likely to stay at the focal point of DfM for as long as optical lithography is used.

3.2.1.1 Static Random Access Memory Cells

Cell-based case studies of pattern fidelity are often carried out on SRAM (static random access memory) cells, which among the many types of standard cells are often considered a showcase for technology capability [16,17]. They are not only one of the most dense, complicated, and technologically advanced building blocks of the ICs, but also the most manufacturable ones, due to the rigorous optimization process they undergo. IC technologies use SRAM matrices for CMOS process monitoring. DRAMs (dynamic random access memory), which provide the highest bit count density, have too complex a front-end process, whereas the key challenge of the EPROMs (erasable programmable read-only memories) is oriented at their reliability related to the gate stack, reducing their usefulness as yield vehicles for generic CMOS process development (Table 3.2). From the standpoint of generic MOSFET

TABLE 3.2

Types of Memories and Their Usefulness for CMOS Process Development

	Access Enabler	Storage Node	Disturbance	Usefulness
NVRAM (EPROM, EEPROM)	High voltage	Gate oxide	High voltage, radiation	Gate oxide optimization
DRAM	MOSFET	Capacitor	Radiation, power OFF	High-density matrices
SRAM	Inverter	Silicon substrate between two MOSFETs	Radiation, power OFF	Device and integration
MRAM	MOSFET	Magnetic momentum	Magnetic field	Magnetic layers

performance, it is important to review DfM challenges in SRAMs, as many of them are applicable to other types of cells across technologies.

One of the reasons for the memory cells to be an example of ultradense packing density is that they follow the key mechanical engineering (ME)/ assembly DfM principle: It is easier to build many complex, but identical, elements than a few simpler ones that differ from each other. The optimization, which would ensure a wide margin for all process aspects (i.e., no hot spots related to reliability, pattern transfer, or parametric data) would have to focus on a limited number of failure modes, preferably not interacting with one another. As one consequence, any given SRAM cell architecture should preferably be used without major rework for several technology generations, that is, be scalable, if not directly by a dial-in procedure, then at least by minor adjustments to the geometries.

The SRAM layout streamlined for robust implementation should have

- simple geometries on critical layers (active, gate, lower-level metals): rectangles print and scale down easier than multiangle polygons
- no acute/reentrant angles
- enough room in the cell area for OPC application to the features that may be hard to print (e.g., minimum pitch has to be expanded for bent lines)
- minimum number of contacts as required to enable enough room for assist features around each of them
- noncritical layers (such as the implants) drawn at large CDs and tolerances so as not to become critical for the next technology generation and compromise layout scalability should not be compromised

One approach to qualify pattern transfer techniques for dense layouts of multitransistor memory cells such as SRAMs is to use integrated optical and electrical simulation to calibrate the effect of pattern reproduction on electrical parameters [18]. This approach is a high-accuracy die level option compared to the local check of EPEs, which do not directly represent the acceptable margins of design intent.

The correlation between device geometries and electrical performance can be verified by comparing predefined control lines, called *tolerance contours* or *bands* with a simulated silicon image of the layout, including the resolution enhancement features. Designs manufacturable within the required process window would have the image lines within those bands. From the device modeling standpoint, tolerance bands may be represented by drawn geometries, upsized and downsized (e.g., by 10%), but this definition may be difficult to match all lithographic pattern distortions. Tolerance contours for lithography may instead be based on simulated and electrically verified aerial images for light intensity thresholds, retrofitted back into layout database. For transistors and interconnecting layers, verifying that such shapes

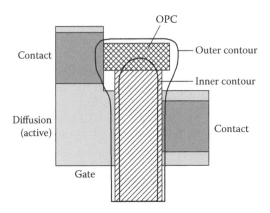

FIGURE 3.3
Example of tolerance contours for a MOSFET in SRAM cell.

are aligned with device models would require SPICE simulations. As an example, tolerance contours for SRAM cell can be used to validate the scalability of critical areas.

The layout representing target silicon geometries supplemented with two contours, corresponding to the inner (minimum) and outer (maximum) CD tolerance lines (Figure 3.3) can be compared to the simulated aerial images with defocus and distortions (flare, aberrations, mask error enhancement factor [MEEF], etc. [19,20]). For metal oxide semiconductor (MOS) transistors, the maximum and minimum lines would control drive and leakage currents, I_{ON} and I_{OFF}, respectively. The shapes of optically imaged MOSFET channels may be far from the desired rectangular shape, with CDs varying locally by more than ±10% across their width, for example, at line end enclosures or overlaps. Because gate overlay to the channel also has an impact on the I_{ON} and I_{OFF}, tolerance contours should also reflect the possible channel distortion within the 3σ misalignment with respect to the nominal channel placement. The fitting of the silicon image between tolerance contours can be verified by a design rule check (DRC). To accurately simulate geometry-dependent drain current, the transistor channel can be divided into multiple parallel sections [18] with I_{ON} and I_{OFF} for each section integrated later over the channel width, followed by comparing them with the nominal values for the rectangular MOSFET, in line with universal channel length dependence model (Figure 3.4) [21,22].

One advantage of predefining tolerance bands over the EPE analysis is the ability to quickly verify the scalability of cell architecture to the next technology node or to enable extra revenue from a linear shrink with limited scope (few percent with respect to the nominal technology node, as below).

For digital devices, a simple method to increase the number of die per wafer without design effort is optical reduction of the entire mask (layout) database at the mask shop. The dialed-in shrink factor could not be too aggressive as it may invalidate process models, but a revenue increase by over 10%,

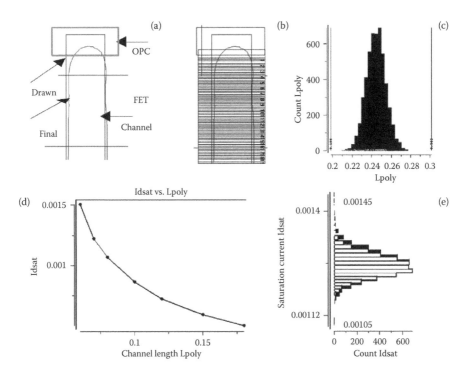

FIGURE 3.4
Integrating procedure of channel length-dependent MOSFET current for simulated silicon image of gate layout: (a) drawn geometry; (b) channel division; (c) channel L distribution; (d) current calibration; and (e) current distribution to be integrated.

corresponding to a linear shrink by, for example, 5%, could be achievable without the loss of yield, performance, or reliability. Assuming identical transistor models pre- and postshrink, pattern restrictions for scaled MOSFETs would be represented by the scaled I_{ON} and I_{OFF}, which to the first order can be verified by the adherence to the simulated scaled tolerance contours (Figure 3.5) [22]. If the leakage current related to the inner tolerance contour does not exceed the upper specification (spec) limit before and after shrink, good scalability of both design and OPC is expected. However, because the drive current related to the outer tolerance contour meets the lower spec limit only for the longer device, one needs to ensure that the I_{ON} requirement is also met for the shorter device in one of the following ways:

- by relaxing the outer tolerance contour (corresponding to the derating of the specified product parameters derived from it, e.g., the frequency of operation)
- by changing the optical exposure model (corresponding to the different lithography process conditions, i.e., use a manufacturability for design [MfD] approach)

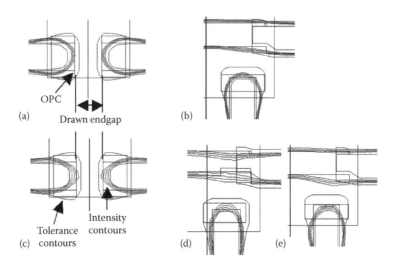

FIGURE 3.5
Cell scalability verification based on tolerance contours. a) before shrunk end cap-to-endcap, b) before shrunk, endcap to line, c) after shrunk, endcap-to-endcap, d) after shrunk, endcap-to-line with OPC (option 1,) e) after shrunk, endcap to line with OPC (option 2).

- by adjusting OPC (i.e., design would change the layout data or CAD would change the OPC algorithm based on the adjustment of the optical model)

To take advantage of the simple optical scaling, device geometries on all design layers should be analyzed in a similar way. For interconnections, tolerance contours would need to prevent opens and shorts (continuous lines must not break, and separate lines must not merge) to ensure that metal bus widths are adequate for EM and that precision resistors have tight CD tolerance.

Shifting contact or via tolerance contours along the interconnecting lines would simulate the effect of misalignment on their enclosures (Figure 3.6).

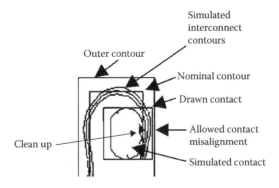

FIGURE 3.6
Verification of overlay tolerance based on simulated aerial images related to tolerance contours.

Maximum tolerance contours for fully enclosed contacts (e.g., hot aluminum) would require minimum DRC space checks between the landing pads. For borderless contacts, a DRC needs to prevent an electrical short to the adjacent connecting lines.

Cell optimization based on simulation and tolerance contours should help increase layout density, especially to verify aggressive OPC (scaling of the k_1 factor in the Rayleigh equation down to very low values). For near-wavelength photolithography, linear shrink of the cell by about 30% (cell area reduction by 50%) may be possible by applying aggressive OPC to scalable layout architecture. Cost reduction by design reuse may also be applicable to periphery circuitry (I/Os, word line drivers) [23].

For deep subwavelength optical lithography, layout scalability depends on the different of types of optical proximities required by the cell architecture [24]. For example, in the DWL (double word line) cell, MOSFETs have only one type of proximity (line end to line end), while in the SWL (single word line) cell, there are two types (line to line end and line end to line end) (Figure 3.7). When the DWL cell is linearly shrunk, the exposure conditions would be optimized only for one type of OPC to ensure a manufacturable process. In contrast, for the SWL cell, OPC and photo process adjustments are likely to be in conflict between the two types of proximity for the image to fit within tolerance bands. For multiple types of proximity, related to the variety of line widths and spacings, it is difficult to propose layout adjustments working simultaneously for all of them and enable footprint reduction by optical scaling. Therefore, the cells featuring many types of proximities are considered poorly manufacturable, and only an aggressive DfM approach can help take advantage of their smaller footprint if they compromise the process window [25].

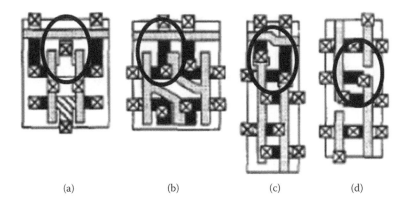

(a)   (b)   (c)   (d)

FIGURE 3.7
Architecture-dependent types of line proximity in SRAM cells. (circled): a) line end to line (single type), b) line end to line (mixed types), c) contact enclosure to line, d) line end to line end. (From M. Ishida et al. 1999. A novel 6T-SRAM cell technology designed with rectangular patterns scalable beyond 0.18 μm generation and desirable for ultra high speed operation. *Proceedings of IEDM*, 201–204, December 1999. With permission.)

3.2.1.2 Stitch Cells

In addition to the efforts focused on the memory matrices, DfM for SRAM and other memory cells has to account for the problems related to the transitional layout regions of stitch (border) cells, which enable connectivity transfer from core to random logic. Most IC products use multiple-layout environments with pattern properties varying among the blocks in terms of pattern density or CD distribution. Memory blocks or image sensors integrated with core logic require different layout enhancement methodologies.

To ensure that the transition of layout properties between the different regions is "smooth" enough not to cause yield loss, the region of stitch cells may need to be designed with the depth of multiple cell pitches. To reduce the risk for reliability, process variation, and model accuracy hot spots, one can

- add dummy rows/columns of core cells modified to be electrically inactive
- develop dedicated a dedicated set of rules for the stitch cells
- define dedicated OPC targets for transition regions
- place a dummy pattern randomizing the interface of "structured" cell matrix with the "amorphous" region of the periphery logic

Usually, periphery rules alone cannot be applied to the stitch cell region due to the vicinity of the core. Transitional features may have unfavorable shapes: Straight lines usually do not allow matching the tight pitch in the core to the relaxed pitch in the random logic. DfM-unfriendly geometries such as L-kinks, protrusions, and telescopic shapes, which need custom OPC, may

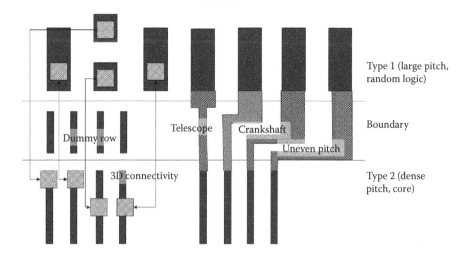

FIGURE 3.8
Challenges of stitch cells at design rule boundaries telescopes, crankshafts, etc., all causing OPC problems. Connectivity, may be enabled by 3D structures.

be required (Figure 3.8). In addition, testing the functionality of stitch cells may need custom procedures if they contain no accessible devices.

To ensure that stitch cells are not a yield limiter, one may define and follow as detailed physical (optical) and electrical models for them as for the core cells. It may appear that when compared to the core cells, the return on design and technology time investment for the stitch cells is much lower because they only correspond to a small fraction of the IC area and usually perform no special function. In addition, unlike the core cells, border cells are often application specific; their development may take a few cycles of process optimization or design training for each type of product. At the same time, a natural approach to "play it safe" (i.e., make the border cells relatively large, allowing ample room for phasing out pattern proximity, device connectivity, etc.) may not work as large geometries may not interface well with the small ones. To better identify hot spots in border cells by simulation, they should not be placed in sensitive areas of the die (e.g., close to die corners) due to their unpredictable responses to process variations.

3.3 DfM at Block and Die Levels: Rules and Best Practices

While development of core and stitch cells requires dedicated simulations, a high-quality layout of random logic at the block and die levels should be sufficiently controlled by design rules, which guarantee that, regardless of the context, they ensure 100% yield. However, due to the gap between the intended and the actual layout shapes, design rule verification becomes context specific, making it necessary to divide the rule deck into three categories (Table 3.3):

TABLE 3.3

Design Rule Categories

Rule Category	Explanation	Example	EDA Complexity
Mandatory	Ensure basic product functionality Valid for any layout context May be more aggressive in fixed environment (memory core)	Width Spacing Enclosure (W/S/E)	Simple (<1 man-day per rule)
Recommended	Introduced to improve yield No impact on individual device functionality but on their variability	Notch reduction, device matching	Moderate to high (1 man-day to 1 man-week)
Best practices	Reduce variability for multiple products in the same way, align parametric deliverables	Via doubling	Coding may not be recommended due to ambiguous solutions

- "Mandatory" rules (DRC). These rules have to be observed regardless of the context; they pertain to the immediate interaction of geometries (i.e., within α times minimum feature size, where α is in the range from approximate 1 to about 3).
- "Recommended" rules (RRs), which act as a precaution against contextual defects (hot spots) causing low yield. They should prevent undesired distribution of layout geometries in the medium interaction range (i.e., for $3 < \alpha < 10$).
- Best practices, which help avoid potentially compromising contexts, usually at the die architecture level (long range) and are too complex to be coded into DRC desks due to the large number of contextual geometries, typically for $\alpha \geq 10$.

As the range of interaction α increases, the function of design rules or best practices shifts from preventing pattern defects ($\alpha \approx 1$) to improving the process window ($\alpha \approx 3$) and ensuring predictable functionality ($\alpha > 10$).

Product diversification and geometric sensitivity due to device scaling are the reasons for the variability of device parameters (threshold voltage V_{th}, local power supply V_{dd}, drive current I_{on}) becoming a challenge to IC functionality and yield [26]. While building robust, but identical, cells no more suffices for the marketplace, products with a high content of diversified, analog circuitry (SoC or analog/RF) driving new market applications still need to take advantage of a common technology platform, which could have been developed based on a low-defectivity SRAM process. These products, similar to SRAM cells, require minimal on-die device-to-device variation (e.g., to match input resistances of differential amplifiers), but in contrast to their SRAM counterparts, their exact reproducibility is no longer ensured by the layout context. Intentionally identical devices may only be provided by an elaborate system of design rules for device matching and placement, protected by a dummy environment of stitch cells, as required.

Medium- and long-range pattern interaction rules may require alternate approaches. Parametric variability of analog ICs is due to both process and layout variations and can also be minimized by both process and layout solutions. Process improvements (MfD), related mostly to equipment upgrades, help avoid delays due to implementation of the matching rules, layout rework for parasitic extraction, timing closure, and potential die footprint increase. Despite the high cost of new equipment, the criterion to choose between DfM and MfD is mostly product revenue loss due to delay to market, which should be built into the return-on-investment (RoI) analysis (see Chapter 4).

Design optimization to mitigate the impact of process variability can be accomplished at several stages of design flow. The best approach is to do it in the form of CBC cells at die definition by involving standard geometries with precharacterized layout. To avoid hot spots caused by arbitrary block placement, one should control layout interaction effects at all ranges and use sufficient guard bands to isolate intentionally identical layout blocks from their environment to ensure reproducible device properties.

If DfM is not built-in up front, drastic measures such as parameter derating, which would lower the ASP, or an all-mask product revision, at significant cost and schedule pushout, may need to be considered to enable volume production.

3.4 1D, 2D, 3D, 4D Variability Compromises

As discussed, one capability of DfM is to trade design performance for process stability. Variability is a key metric of success of stable DfM at a high parametric level. To keep track of it, design rules to control the impact of layout and process variability on product performance should be defined in four dimensions (Table 3.4).

DfM for variability reduction in one or two dimensions should involve as few CAD corrections as possible, as they work against predictability and design stability (i.e., can produce unexpected results). Extractable CBC design in three dimensions is more robust than one touched up by containment CAD solutions. At the same time, design may be required to prioritize between layout improvements for different types of hot spots, and the fixes may be inconsistent, depending on their goal. Particle defect reduction through CAA, LfD, and for CMP limited yield (CMPLY) improvement demand multiparameter layout optimization, and their joint consequences on the final yield data may be difficult to predict, although studies show limited impact of the sequence of layout corrections for the different process modules on the final yield.

TABLE 3.4

DfM Development in 1D–4D

Dimension	Key Issues and Approaches
1D	Printability of long lines at fixed pitch as a function of process variability
	Needs conservative L/S pitch and dummy environment
2D	Corrections for line width transitions impacted by layout and process variability
	Design rules for dummy environment
3D	Process/layout interactions impacted by die topography
	Pattern density resolution and coupling capacitance rules
4D	Time-dependent performance related to device and interconnect wear out
	Impacted by interactions among process/layout/product function

TABLE 3.5

Basic MOSFET Parameters and Sources of Their Variability

Electrical Parameters	Examples of Process Parameters Impacting High Variability
Drive current	Implant distribution, contact enclosures, sheet resistance
Leakage current	Implant-related electric field at diffusion edges
Breakdown voltage	Doping distribution, thermal process, line spacing (horizontal and vertical)
Contact resistance	Metal enclosures, liner adhesion, thermal processes
Line resistance	Metal grain structure, line edge roughness
Capacitance	3D effects, poly depletion, ILD planarity, pinholes

For the fourth dimension (i.e., in time or voltage domain) in which the IC has to operate, design rules should guarantee 100% functionality and reliability within the assumed process variability margin (e.g., 10%). The process excursions translate into parametric variations, which may be acceptable for the digital, but too big for analog/RF products (Table 3.5). As an example, variability control of a pattern transfer process requires compensation of mask CD data based on photo and etch bias extracted as a function of layout dimensionality (Table 3.6):

- 1D: pitch bias-dependent CD corrections to lines and spaces
- 2D: OPC adjustments at line width transitions (pullback and rounding)
- 3D: prevention of multilayer effects (pattern density impact on line and contact CDs) but also architectural rules for SoC or SiP
- 4D: time- and voltage-dependent electrical CD variations (e.g., due to hot carrier injection [HCI], negative bias temperature instability [NBTI]),

TABLE 3.6

Variability Reduction Options Pros and Cons

Option/Responsible Party	Design (DfM)	CAD	Fab Process (MfD)
Local fix of layout defect	Easy but limited in scope, may need to be repeated multiple times	May create unpredictable changes of layout	Unlikely to address only the root cause
Reengineering of design or fab flow	Time-consuming model redevelopment	Unclear scope	High cost of new process setup
Derating of high-variability devices	Not recommended, may increase die footprint or lower the ASP	Verification algorithms easy to define but limited in scope	Tuning of devices to compensate the problem (e.g., by implant changes)

Because standard, short-range DRC is not able to ensure low variability, approaches based on RR and best practices are required for multidimensional problems, as discussed next.

3.4.1 1D: Line/Space Pitch

The 1D variability on a wafer is the smallest for an "infinite" set of "infinitely" long, straight lines printed at a fixed pitch, which can be used, for example, to create transistor matrices. Because in reality straight lines cannot extend indefinitely in any direction, perimeters of such matrices should be terminated with dummy lines to protect active devices from proximity effects due to the adjacent circuits and to control the CD variation due to proximity. If design intends to reduce process-related variability within such pseudoinfinite matrices, the low-variability rules should allow minimum CDs of the lines to be larger than minimum CDs based on technology capability:

$$CD = \min CD_{Drawn} \cdot VM \tag{3.3}$$

and

$$\min CD_{Drawn(L)} = \min CD_{Drawn(S)} \cdot \frac{SVM}{LVM} \tag{3.4}$$

Parameter	Units	Explanation
ΔCD	Length units	CD variation based on process capability
$\min CD_{Drawn(S,L)}$	Length units	Minimum CD of drawn lines (S is a standard process, and L is a low-variability process)
VM	Ratio (%)	Allowed variability margin = $\Delta CD/CD$
SVM	Percentage	Variability margin for standard process capability (e.g., 10%)
LVM	Percentage	Variability margin for low-variability process (e.g., 2%)

The intrinsic (i.e., non-design-related) CD variation in 1D is caused by the line edge roughness (LER), exposure latitude, and mask CD variation amplified by the MEEF. To achieve the desired LVM at fixed pitch, one needs to selectively increase minimum CD layout geometries by the SVM/LVM ratio (e.g., 10%/2%, i.e., by 5×) or apply process bias compensation by CAD sizing for the line/space ratio to be close to the optimal value (e.g., 1:2 is considered robust by photo engineering). In general, 1D pattern variations on a wafer are caused by material properties, and their correlation to design/DfM modifications may be poor.

3.4.2 2D: Pattern Enhancement Techniques

The first requirement of 2D variability reduction is high resolution of pattern transfer from design to wafer [25]. Since the beginning of IC manufacturing, this process was based on photo imaging, with wafer-level resolution defined by the formula proposed by Rayleigh in the 1800s for a diffraction-limited projector:

$$R = k_1 * \lambda/NA \tag{3.5}$$

Parameter	Units	Explanation
R	Length units	Minimum width of resolvable line
λ	Length units	Wavelength of the imaging light
NA	Unitless	Numerical aperture of the projection optics
k_1	Unitless	Resolution factor depending on the process conditions and aggressiveness of layout corrections (OPC) for half-pitch or single lines

The basic width/spacing/enclosure (W/S/E) rules ensure the accuracy of pattern definition within few a tens of nanometers of the feature to which they apply. The rules for low variability must include the impact of medium- and long-range pattern interactions (across the IP block, die, or exposure field). Verifying layout printability depending on die architecture requires first defining the rules for power routing, especially if the routing layers for different packaging options are a part of that architecture.

While CBC reduces the number of random layout geometries, which may contribute to hot spots (Figure 3.9) [26], it may be difficult to implement at higher levels of die integration due to product-specific IP content and metal routing. And, even for the parameterized CBC layouts, the parasitics are still

Design Feature   Random Defect

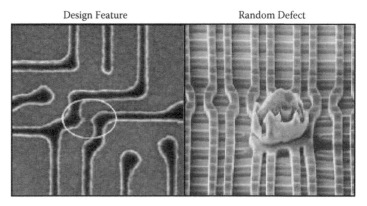

FIGURE 3.9
Comparison of a 2D wafer image design of a hot spot to a random defect. (From Y. Shim et al. 2008. Improvement on OPC completeness through pre-OPC hot spot dection and fix. *SPIE*, vol. 6925, p. 692513. With permission.)

context-sensitive, which can limit flexibility of their placement, without prior extraction.

In two dimensions, ΔCD depends on the imaging process window related to exposure and focus variations, lens aberrations, etch bias, and so on, and can be divided into two components:

$$\Delta CD_{(x,y)} = \Delta CD_{\text{Img}(x,y)} + \Delta CD_{\text{Other}(x,y)} \tag{3.6}$$

EDA tools can extract the $\Delta CD_{\text{Img}(x,y)}$ based on aerial images from calibrated lithography models. Then, ΔCD_{Other} becomes

$$\Delta CD_{\text{Other}(x,y)} = CD_{\text{Drawn}(x,y)} * VM - \Delta CD_{\text{Photo(simulated)}(x,y)} \tag{3.7}$$

Parameter	Explanation
$\Delta CD_{\text{Img}(x,y)}$	Local CD variation due to the imaging process (photo)
$\Delta CD_{\text{Other}(x,y)}$	Local CD variation due to other process effects (e.g., etch)
$\Delta CD_{\text{Img(simulated)}(x,y)}$	Simulated local CD variation at photo

To the first order, $\Delta CD_{\text{Other}(x,y)}$ in 2D can be assumed to achieve maximum value for the 1D layout in both x and y directions:

$$\Delta CD_{\text{Other}} = \Delta CD_{\text{Other}|1D} \tag{3.8}$$

For a 2D layout, ΔCD_{Img} is extracted from simulation:

$$\Delta CD_{\text{Img}|2D} = \Delta CD_{\text{Img(simulated)}} \tag{3.9}$$

As in the 1D case, the question is by how much would the minimum $CD_{\text{Draw(L)}}$ be larger than the minimum $CD_{\text{Drawn(s)}}$. The answer would depend on the layout context driving the SVM/LVM ratio. Typically, for geometries other than straight lines, acceptable SVM can be much larger than 10% if local shape distortion is inconsequential to device operation because design rules prevent placing devices in the corners of drawn geometries. But, the same may not be acceptable for LVM due to the sensitivity to the parasitics. LVM can then be calculated from

$$LVM = (\Delta CD_{\text{Other}|L1D} + \Delta CD_{\text{Photo(simulated)}})/\Delta CD_{\text{LDrawn}} \tag{3.10}$$

All the components of this equation need to be obtained from simulation and extraction of 2D calibration structures for the imaging process. Basic 2D effects correspond, for example, to the change of the line width, from CD1 to CD2 (Figure 3.10 [27]) where CD2 may be 0 for line termination, semi-infinite for change of line direction, or a value different from CD1 for a telescopic joint.

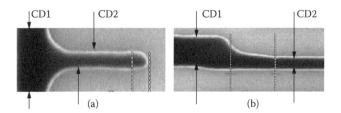

FIGURE 3.10
Regions of line width transition from CD1 to CD2 (MOSFET gate region) which need to be spaced away from the MOSFET gates.

For the 2D variability scenarios, layout ΔCD would inherently be larger than it is for 1D. To stabilize ΔCD and extract the parasitics, the lithography may require large transition regions around the transistor gates (Figure 3.10).

- minimum extension of poly beyond active (endcap)
- minimum spacing of the gate to the inside or outside active corner
- minimum spacing of the gate to the line of extended width

As a result, the area required by a device may grow by multiple times (e.g., 2× minimum CD + misalignment) to ensure low variability in its critical, active region. LER from 1D extraction should also be taken into account (e.g., by increasing the area-to-perimeter ratio). MOSFET channels have to be free of the pullback and rounding due to insufficient OPC on the one hand and of CD ringing due to aggressive OPC on the other. The endcap problem may be solved by the "gate-first" process, but other printability and pattern resolution issues would remain.

Improving wafer planarity is necessary to reduce 2D variability within device-related layers due to the depth of focus having an impact on their ΔCD. One way to ensure it is by adding dummy structures at the perimeter, within an experimentally derived distance DRW of the layout matrix (Figure 3.11):

$$DRW = \alpha * \lambda * UF \qquad (3.11)$$

Parameter	Units	Explanation
DRW	Length units	Minimum width of a dummy cell or ring
UF	Ratio-unitless	Uniformity factor equal to the desired $\Delta CD/CD$ ratio
λ	Length units	Wavelength of the imaging light
α	Unitless	Proximity sensitivity factor, proposed $\alpha = 4$

3.4.2.1 OPC

Rule-based (RB) and model-based (MB) OPC are the key techniques to improve pattern resolution by reducing k_1 [28]. Because standard design extraction

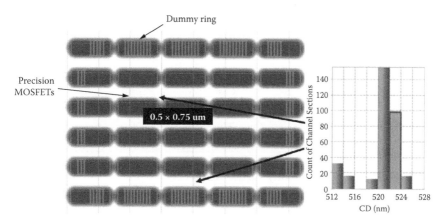

FIGURE 3.11

Precision MOSFET matrix with a dummy ring ensuring tight CD distribution. Total distribution (nm) around the minimal value of 520 nm. (Reprinted with permission.) The number of bars corresponds to the number of channel sections with L different from nominal value by more than 2 nm.

procedures are limited to straight lines terminated by 90 deg corners, it is desired that wafer pattern also consists of such features. Rule-and model-based and OPC both compensate for multiple process effects potentially preventing loss of design intent. Wafer pattern, as-printed, is not "intrinsically wrong," and its corrections only modify problematic features causing mismatch with design intent. Here, the impact of RB touchups to layout can be confounded by multiple side effects. One cannot distinguish whether the sizing of a MOSFET is required by the process bias, model inaccuracy, or extra process margin and how to modify it for different devices or manufacturing lines. They also work typically for only one manufacturing environment or tool set.

As a DfM technique, OPC comes with its set of options, costs, and associated RoIs (Tables 3.7 and 3.8). The examples show that RoI on OPC is significant multiple tens of percentage prints.

TABLE 3.7

RoI for OPC-Examples of Calculations Assuming a 3-year Product Development Cycle

ASP Justification of OPC			ASP Advantage
ASP benefit at launch	Increased GDPW	200%	A
	New applications	150%	B
	Total	300%	$A \times B$
ASP erosion rate	Per year	10%	C
Time to ramp	Years	3	D
ASP erosion from launch	% total value	27%	$E = 100 - (100 - C)^D$
ASP after erosion		220%	$A \times B \times E$
Cost-benefit comparison	170% cost	220% benefit	

See Table 3.9 for OPC cost calculation.

TABLE 3.8

OPC Cost and Area Advantage Depending on the Option (Yield Improvement is Not Accounted For)

OPC	Relative Layout Area	Cost	Example Benefit	Benefit/Cost Ratio
Aggressive	50	150	200	1.33
Just enough	80	130	125	1.05
Minimalistic	100	100	100	1.00

RB OPC is applicable mostly for 1D layout corrections at different pitches and with limited 2D pattern enhancement capabilities, such as corrections to line end extensions (LEEs) or corner rounding. Sometimes referred to as "sizing" or "process bias compensation," RB OPC is a catchall routine for the numerous layout-to-wafer mismatch problems. It can help

- mitigate the effect of nonuniform line pitch or pattern density on photo or etch microloading
- compensate for lateral spread of dopant (e.g., for well implants)
- selectively change device geometries (e.g., to improve high-voltage performance)
- mitigate the impact of corner rounding or line end pullback by adding LEEs

When increasing contact enclosures using LEE applied to metal (Figure 3.12), RB OPC can be tuned down such that a "ringing" effect is not created, unlike the more aggressive option of MB OPC, making RB OPC less sensitive to process variations. As a result, the MEEF for RB OPC would be close to 1.

By contrast, MB OPC, a workhorse of resolution enhancement for technology nodes from 130 nm down has been successful in reducing k_1 to below 0.3, as a conceptual extension of RB OPC at a higher complexity level, but with much higher sensitivity to both process variations and layout context. While RB OPC shifts the edges of layout geometries by upsizing or downsizing to improve proximity conditions, MB OPC would first partition these edges and then apply individual corrections to each portion based on the local model parameters. Depending on the intended accuracy, the complexity of the shifts would vary.

MB OPC has many disadvantages, such as

- higher cost compared to RB OPC due to long development cycle time and complexity (Table 3.7)
- large size of the database and long mask write time
- potential for creating uninspectable features due to the layout complexity

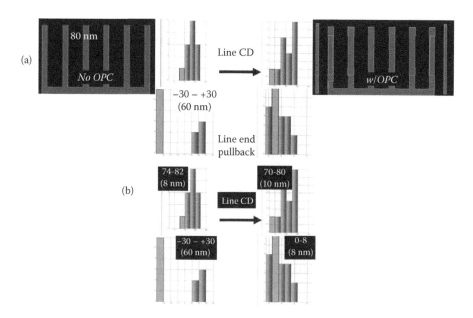

FIGURE 3.12

CD distributions for geometries printed without and with OPC. Comb structure: (a) drawn and mask layer with model-based OPC; (b) the resulting CD distributions showing pullback reduction from 30 to 8 nm at the expense of "OPC ringing" of 4 versus 5 nm at the MOSFET channel.

Substituting RB with MB OPC is related to both change of the partitioning methodology itself and the upgrades to the process and design architecture. For example, the onset of MB OPC was concurrent with replacing the LOCOS isolation process by STI, which reduced the spacing between two active areas of opposite doping types (N+/P+) and enabled the shrinkability of the rules related to the photo process to drive a change in the SRAM cell architecture which in turn required aggressive OPC to compensate for the layout complexity.

Resolution enhancement techniques (RETs) requiring custom OPC entered production at the 180-nm node to modify physical properties of a wave front on the mask. They enabled control of wave direction by off-axis illumination (OAI), control of wavefront amplitude by aperture sizes and shapes, and control of the local wavefront phase by material transparency or refraction of the phase-shifting masks (PSMs) [29].

Two popular combinations of RET are OAI with subresolution assist features (SRAFs) for gate layers and attenuated phase shifting with OPC for contact layers. For OAI, the illumination falls on the mask at angles resonant with the pitch of periodic structures in the layout (ideal for IC arrays such as DRAM). The on-axis components of the image, which do not add contrast, are reduced or eliminated (Figure 3.13).

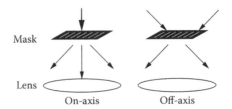

FIGURE 3.13
On- and off-axis components of the image. (From F. M. Schellenberg et al. 2001. Proceedings of Design Automation Conference. p. 89. With permission.)

The application of the geometric configuration of OAI (dipole, quadrupole, or annular illumination [30]) depends on the type of pattern to be printed because only certain pitches and periodic patterns are enhanced, while others are compromised. For example, in quasar illumination four sections of an annulus form four poles of light falling on the mask from four orientations at ±45°. This provides excellent enhancement for Manhattan geometries, but 45° orientations scatter this light poorly and may not be imaged on the wafer at all, leading to highly constrained design rules. A twofold enhancement is possible for small, dense pitches using quasar illumination (Figure 3.14).

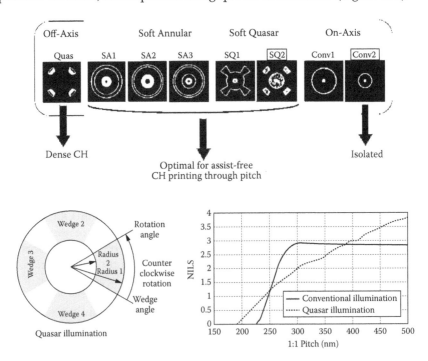

FIGURE 3.14
Lithography light source options: quasar and conventional. NILS-normalized intensity log slope annular.

However, the iso dense bias problem [31] arising when features on the mask with the same line width print on the wafer with different dimensions can be amplified with OAI. The angle of the light leaving the mask depends on the pitch of the mask structures, and diffracted light at higher angles (smaller pitches) can be attenuated. This leads to the bias effects between isolated and dense lines. Because the OAI technique enhances the diffraction to the outer portions of the lens, the impact to isolated lines can be even more severe and can give rise to "forbidden pitches," in which a line of a particular feature size would be prohibited from being placed at certain spacing to other features [32]. Compensation for the iso dense bias can be achieved using an inverse bias, applied as a design rule over the layout. However, the isolated and dense lines diffract differently in the lens, and various imaging properties (such as depth of focus) can still be different. By placing small additional features (SRAFs) on the mask near isolated or semi-isolated lines, the diffraction pattern from the mask becomes close to that of a dense line. Similar parts of the lens are used, similar transfer properties are observed, and the iso dense bias is reduced.

But, the addition of SRAFs to improve small feature resolution comes at additional cost to the mask and pattern verification. The layout of the mask no longer resembles the intended designed layout (Figure 3.15), and a DRC would need to be modified. The distinction between the RET and designer's intent features must be provided by design flow, typically by using different CAD layers combined after separate verification into a common mask database.

In contrast to the simple rules that govern the placement of SRAFs, MB OPC represents a very different style of RET [33]. In MB OPC, the flow for generation of the mask layout itself contains a process simulation step that predicts the final result on the wafer and adapts the layout until satisfactory convergence between the final image and the original physical design intent is achieved. This is typically inserted as part of the design verification.

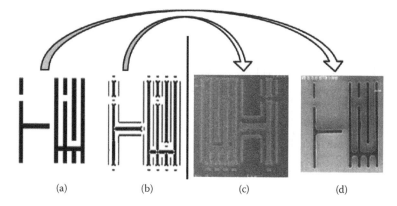

(a)　　　　　(b)　　　　　(c)　　　　　(d)

FIGURE 3.15
A comparison of drawn data, mask data mask, and silicon image: Mask data (b) adding SRAFs and serifs (c) to ensure a close match of (a) and (d). (From F. M. Schellenberg et al. 2001. Proceedings of Design Automation Conference, p. 89. With permission.)

For imaging layers with small contact holes, the production of arrays is executed using masks with attenuated phase-shifting layers in which the opaque mask material (typically chrome), replaced with a slightly transparent material that also causes a uniform phase shift of 180°, passes only enough light to cause a dark interference fringe to form at the boundaries of the features. This enhances image contrast and image integrity but typically does not allow enough light to pass through the mask to actually expose the photoresist in the dark areas. Low light levels transmitted through multiple regions can cause some undesired artifacts on the wafer, when constructive interference from adjacent contacts occurs, and an additional bright spot over the exposure threshold forms on the wafer. The circuit design does not include these additional features, called "side lobes," which can have a severe impact on the device yield (Figure 3.16).

To remove additional light from the areas where side lobes form, one needs to add an opaque chrome patch to the mask in an additional patterning step using common patterning techniques and three-layer mask blanks. The layout software must first identify the locations where to insert these blocking structures, which can be complicated, when the areas of the contact apertures are being dynamically modified by MB OPC programs. An additional check for MB OPC simultaneously alters the aperture shapes for improved fidelity. When a side lobe is detected, a patch of chrome to cover it is generated, and the OPC continues with the chrome patch in place. The final reticle layout will contain three layers, one for attenuated material, one for clear areas, and a third for chrome patches ("tritone" mask).

Unlike for SRAFs, attenuated phase-shifting mask blanks are available at about twice the cost of the COG (chrome on glass) mask. The final cost of ownership for this technique commonly used for contact and local interconnect

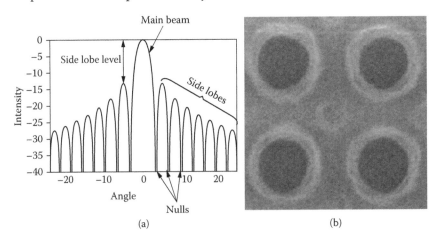

(a)   (b)

FIGURE 3.16
Side lobes in silicon image. (a) intensity contours; (b) top wafer view. (From F. M. Schellenberg et al. 2001. Proceedings of Design Automation Conference, p. 89. With permission.)

TABLE 3.9

Aggressive OPC Can Cost more than 170% of Standard OPC for New Product Development Due to the Extra Expenses for Data, Mask, Fab Processing, and Verification

Cost Components of Aggressive OPC (%)	Design Data Processing	Mask	Fab Processing	Metrology/ Verification	Combined
Incremental cost	150	120	200	200	N/A
Total cost of new product launch	10	30	50	10	(100)
Total additional cost	15	36	100	20	

layers is much smaller than for the SRAF technique and is far more widely adopted. However, as with the SRAF technique, the mask layout is again dramatically different from the designer's intent. The layout verification must have a reference to the "target" layer representing design intent for the silicon to enable the wafer image and comparing simulating it to the "target," generating EPE data or tolerance contour checks.

No matter which RET technique is chosen, the differences between the original physical design of the circuit and the layout used to write the mask are now large and growing with each new technology generation. The maintenance cost is also increasing (Table 3.9).

3.4.2.2 Electrical Impact

Because the wavelength of the transverse electromagnetic (light) wave is orthogonal to the plane of the printed pattern, λ can be smaller than on water resolution R by about the factor of k_1 and still provide acceptable pattern quality. Simple OPC already suffices to reduce k_1 to 0.8. Features with CDs $\ll \lambda$ can be printed with phase-shift, off-axis illumination or MB OPC, at k_1 down to below 0.4. But, the key goal of OPC is to preserve device (MOSFET) line width. For critical features located $\alpha \times$ CD away from the line transition regions, printed CDs would be within 10% (SVM) or 2% (LVM) of drawn dimensions. However, OPC is sensitive to process variations and design style, and its advantage depends on trade-offs between the aggressiveness and layout footprint. The lower the ΔCD process variation required, the more k_1 should be increased to improve the lithography process window and the linearity of mask pattern transfer to the silicon. This in turn increases the required guard band around the sensitive geometries, projecting on the layout footprint. MB OPC applied to a comb structure creates a conflict between minimizing the space from the layout singularity to the region of stable line width (i.e., the line end pullback controlled by OPC) and preserving gate CD (Figure 3.12). CD variation across the channel W for the uncorrected lines is typically smaller, while the corrections reduce the line end pullback. For sensitive circuits, that extra

CD variation (which we can call "OPC ringing") would cause risk. Therefore, it is often better to reduce OPC to a minimum, even at the expense of device footprint. The lower k_1 values corresponding to more aggressive OPC translate into a narrower lithography process window, eventually becoming unacceptable from the variability standpoint at $k_1 < 0.25$. For a stable process, k_1 values of 0.6–0.8 are recommended as a rule of thumb for low-variability layouts corresponding to the VMS/VML ratio of about 4 and allowing staying comfortably above the minimum process capability limit. The cost of area increase would have to be justified by the product ASP.

But, even a carefully drawn and fully symmetrical analog layout can potentially be affected by lens aberrations such as astigmatism or coma and fail to deliver the expected product performance. As an MfD response, modern lithographic printing devices (scanners) are equipped with integrated lens interferometers, which allow capturing full-field interferometric data, and measure lens aberrations *"in situ."* Information about field uniformity could be passed to designers such that they place the most distortion-prone geometries (such as analog/digital converters, ADCs) into "clean" imaging field areas with minimum RMS (root mean square) wave-front errors (aberrations).

Upsizing the layout footprint to reduce proximity effects (Figure 3.17) is not the best option to reduce parametric variability of the devices due to the different ranges of layout interactions. To verify whether increasing the device

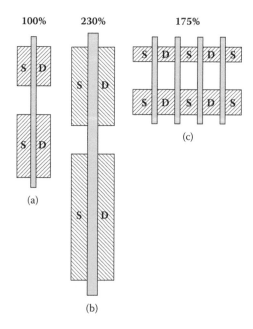

FIGURE 3.17
Layout options of an inverter with the same DC drive current. (a) reference; (b) linear increase; and (c) folded gates. Area increase 5a to 5b, 130% (100% to 230%) (5a to 5c) (100% to 175%). 75%. (From A. Balasinski. 2007. *Journal of Micro/Nanolithography, MEMS, and MOEMS, 6 (03)*. With permission.)

area would improve the yield bears significant cost [34]. Detecting a yield difference of 5% would require at least one split lot. Therefore, die area increases on the order of a few percent are only acceptable if the expected yield benefit is more significant than that. Note that the net die per wafer (NDPW) count would degrade more rapidly due to marginal process capabilities when the die is too small than due to the lower gross die per wafer (GDPW), when the die is too large. In addition, parasitic RLC components for larger devices may compromise product functionality beyond the business impact of the footprint. As an example, three inverters shown in Figure 3.17, identical from the direct current (DC) standpoint, would have different frequency responses. One can trade lithography yield improvement (Figure 3.17a to 3.17b) for point defect yield degradation (Figure 3.17a to 3.17c) or for improved parasitics (folded gates reduce RC). Even if DC simulations show the superiority of the layout in Figure 3.17b over Figure 3.17a [35], the product line designers prefer the parasitics, not just the printability, to be in line with expectation.

The DfM approach to layout optimization is also a function of its ownership among design, CAD, and manufacturing teams (Figure 3.18). The differences in DfM priorities of these teams have the potential to create layout hot spots, random symmetry, and poor matching. Most design rule criteria are inherited from previous technology generations, while the rule values are adjusted for process upgrades and business expectations. The conversion of logical and electrical design representation into layout has many forms—pCells, standard cells, hand drawings, automated routing—but architectural supervision is not always consistent. At the product level, the key goal

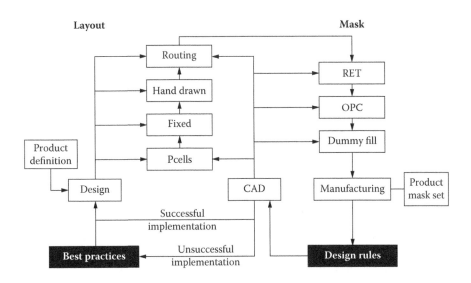

FIGURE 3.18
Feedback loops in design flow among the parties responsible for converting IC design and layout into a mask pattern.

is to achieve timing closure after extraction of all parasitics. However, the feedback from manufacturing back to design may not prevent performance surprises as the paths from problems to solutions are confounded by the multiple process steps in the flow. As an example, the postprocessing of layout into mask data to ensure horizontal (line) and vertical (stack) CD control with OPC and fill pattern may affect the parasitics, but it would not be retrofitted to the design database for extraction and verification. To mitigate this concern, CBC pCells may be optimized for 2D and 3D effects (pattern density and printability). Low-variability rules driven by layout parameterization may be developed for MOSFETs (Table 3.5) to reduce their sensitivity to lens aberrations of the litho system, such as by requiring relatively high $k_1 =$ 0.6–0.7 (at "half pitch" or "CD").

Extraction of CDs for all devices and interconnects enables advanced timing analysis based on (post-OPC) mask data. To provide correct estimates of on-silicon performance, the extraction of residual EPE errors from placed and routed full-chip layouts should provide CD values calibrated to silicon that are then used in timing analysis and speed path characterization. Traditional design flow uses ideal (i.e., drawn) gate length values, leading to a limited accuracy. Improving OPC quality should focus on critical gates or matching transistors for more realistic timing analysis, as it may lead to significant reordering of speed path criticality and an increase in worst-case slack (e.g., 36.4% [36]).

While RETs such as phase-shift masks and off-axis illumination allow for continuous CD reduction, distortions of silicon images are introduced due to proximity effects. Complex and costly OPC is deployed to compensate for MOSFET gate length (L_{gate}) variations. With these modifications, chip manufacturability and yield are improved, but performance may be adversely affected. To consider process variations such as gate CD, oxide thickness, metal width and thickness, temperature, voltage, and so on, during circuit performance analysis at the design stage requires proper modeling of the variabilities [35]. The most common approach, typically aimed at speed/frequency prediction, is based on "worst-case scenarios" (corner cases), which assumes all transistors are independent, and hence yields overly pessimistic results [37]. More accurate variability modeling, by considering its sources treated as purely random components, uses a probabilistic framework with effort to accurately model correlations.

More than 50% of L_{gate} variation is due to systematic sources [38], which can be modeled accurately on layout, assuming realistic systematic contributions (mostly due to proximity effects) to the overall process variation. A location-dependent intrachip variability of L_{gate} leads to large circuit path delay variation. Aerial image simulations can account for fixed-layout patterns, expandable to full-chip timing analysis, while library-based OPC will not become context dependent at the full-chip level.

An alternative technique for post-OPC embedded static timing analysis uses extracting residual EPE errors to derive actual (calibrated to silicon) L_{gate}

values. Such flow combines post-OPC layout back-annotation and selective extraction from the global circuit netlist. Pin slacks of critical instances may be worsened on average if more critical cells and cell types become involved on critical paths.

Having located critical layout patterns across the full chip, where OPC could not achieve the desired L_{gate} control, a calibrated OPC algorithm should be applied locally to attain savings on full-chip OPC runtime, such as in designs with matching transistors (e.g., mixed signal or clock generation in microprocessors) as a supplement to multithreshold voltage assignments, or gate sizing.

OPC compensation for CD distortions becomes a major source of systematic variability itself. Systematic intrachip L_{gate} variability is the main cause of speed degradation for large circuits, especially for a large number of critical paths and short logic depths. Designers only look at worst-case speed paths in timing simulations based on ideal L_{gate} values (potentially corner based), which may be misleading in the subwavelength lithography regime. Aerial images of printed features are used to check the effectiveness of OPC and can provide accurate L_{gate} predictions after a layout is complete. Optimally post-OPC timing verification and recalibration of critical paths should follow.

The starting point of the flow is a post-OPC layout on which process CD simulations can theoretically be performed across the entire chip, although this would be very time consuming. Tagged critical gates are then extracted along with the peripheral geometries within a certain interaction distance (i.e., the optical diameter, beyond which geometries have no impact for the given optical lithography process), on a path-by-path basis. Aerial image simulations are done for the critical paths, and the silicon-based CDs are extracted and defined at the center point for each transistor. Each process CD value is identified with the coordinates of the corresponding transistor to map process CDs back to the circuit netlist.

The next step is library recharacterization, where the cell timing library is updated with silicon-based process CDs by creating a location-aware SPICE netlist for each cell. After the process CDs for critical gates are back-annotated, the original library is expanded with recharacterized cells.

A special LVS is performed to extract the location of each transistor within a cell in the library.

The L_{gate} values in the SPICE netlists are updated with the extracted silicon-based process CDs, and SPICE simulations are repeated for timing recharacterizations to account for the impact of systematic L_{gate} variations and the impact of capacitance changes. The recharacterized cells are reintegrated with the cell library for timing analysis. The global chip netlist is updated to map to the expanded cell library, a full-chip timing analysis is performed using commercial tools, and the top critical paths are reported. These critical paths may include cells that were not recharacterized in the previous step (i.e., they are on paths that the original pre-OPC timing analysis did not flag as critical), in which case recharacterization of newly critical cells will be necessary.

The goal was to determine whether the systematic L_{gate} variations introduced by RET/OPC have a significant impact on typical corner performance. If the optical diameter outside which neighboring geometries have no impact is set as 4 μm, several iterations of cell timing recharacterizations may be necessary at a low runtime (hours). In the final speed path report, under 2% of total critical cells have not been recharacterized using their extracted CDs; these cells appear only in paths with lower critical ordering. Only a small subset of all instances (0.5%) is selected for post-OPC CD simulations, for critical cell paths and timing recharacterization for critical cells done in parallel. Most cells on critical paths are simple, such as inverters and MUXes (multiplexers), so that little runtime for cell timing recharacterizations is required. As a result of optimization requiring layout modification, the OPC layer will be changed, but the majority of the layout geometries remain the same.

With critical paths defined as paths with slack 0, they are rank ordered with the most critical ones first. The experimental results show that the silicon-based timing report lists more paths (2.7×) as critical than the ideal L_{gate}-based timing report, with average path slacks worsened by 24.4% (Figure 3.19). The slack distributions for critical paths from the silicon-based timing report and the traditional timing report provide a binary indicator with values +1, meaning that the path appears in both timing reports, and −1 when the specified path only appears in the silicon-based timing report. The y-axis relates

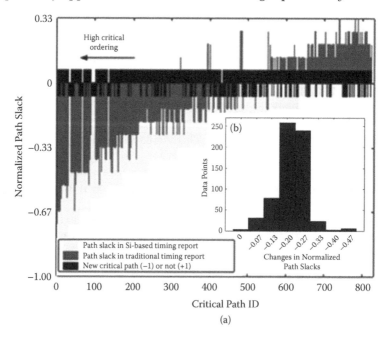

FIGURE 3.19
Impact of OPC on path slack. (From J. Yang et al. 2005. (a) Normalized path slack; (b) (insert) distribution of changes. Proceedings of Design Automation Conference, p. 359. With permission.)

normalized path slack where the slack of the most critical path in the silicon-based timing report is the reference value (there are more negative slack paths in the silicon-based timing report). For paths existing in both timing reports, the sign (negative/positive) of the critical ordering difference indicates the shifting direction (more/less critical, respectively), while the absolute value indicates how many paths it passes to achieve the new critical ordering. (Figure 3.19 indicates 195 paths became more critical and 224 paths less critical.) The path slack changes show the worst-case slack increased by 36.4%, and the average slack change was 0.22 with a maximum shift of 0.47 (normalized to the worst-case slack in the silicon-based timing report). For post-OPC CDs, 21.8% new critical paths are identified, some highly critical with large slack violations. As a result, using the traditional timing analysis, certain paths critical postfabrication will not be considered for resizing.

By identifying the most frequently used critical cells, effort may be placed on optimizing their delay, and better CD control may be achieved with an OPC-friendly layout. Over 50% of critical cells in the timing analysis are made up of the top 10 critical cell types (Figure 3.20). Therefore, library-level design optimizations should be based on a silicon-based analysis.

As discussed, model-based OPC corrects the layout on a point-by-point basis, considering all neighboring features and complex interactions between the stepper and mask. By decomposing the edges of each feature into small fragments, OPC quality can be improved through fine-grain edge movements, which are constrained by neighboring geometries for each fragment. The use of a prioritization scheme for each fragment may limit the convergence of the OPC correction. For instance, the number of iterations of edge movements required to converge below the specified residual EPE error might vary from fragment to fragment and require adjusting the priority scheme based on the identification of the problematic layout patterns. There are several cells with both large CD errors and a high frequency of occurrence, calling for

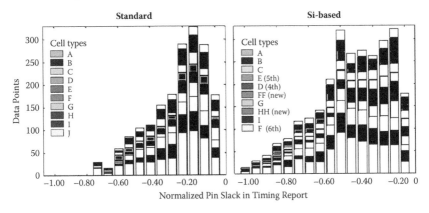

FIGURE 3.20
The difference between standard and silicon L_{gate}-based timing report for critical cell types. (From J. Yang et al. 2005. Proceeding of Design Automation Conference, p. 359. With permission.)

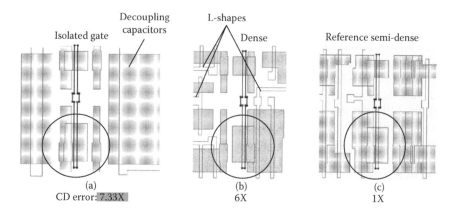

FIGURE 3.21
Environmental impact on L gate residuals in an inverter. (Devices of interest are circled) (a) maximum error for isolated gate; (b) high error for dense gate; (c) reference environment (no error). (From J. Yang et al. 2005. Proceedings of Design Automation Conference, p. 359. With permission.)

either customized OPC recipes or adjusting cell layout patterns to improve printability.

The impact of neighboring features may cause gates to exhibit large CD variations even after OPC. The impact of environment for three instances of a single (identical) gate polygon in terms of L_{gate} errors (Figure 3.21) for an inverter can focus on the P-channel transistor (bottom device), with two heuristic rules to enhance manufacturability: (1) use a single pitch on the critical layer and (2) avoid nonrectilinear shapes (e.g., L's or T's). The layout in Figure 3.21a is undesirable due to the large decoupling capacitances nearby and in Figure 3.21b is poor due to the many L shapes. For the same inverter layout in Figure 3.21c, the neighborhood around the P-channel transistor leads to better printability, reducing the L_{gate} errors by 7.33x compared to Figure 3.21a.

In conclusion, automated flows for post-OPC performance verification should be expected to show differences in the timing analysis compared to the traditional methodology of using nominal, drawn geometries to compare with silicon. The number of critical paths can increase almost twofold, while the worst-case slack violation increases by over 30%. Many critical paths in the post-OPC flow were not reported in the traditional timing analysis.

In the new flow, the post-OPC CD extraction is performed at the same stage as the interconnect parasitics are extracted and back-annotated. In this way, gate CD variations due to RET/OPC are taken into account during static timing analysis to achieve more accurate performance predictions. The methodology requires tagging gates on critical paths or matching them for specific corrections.

3.4.2.3 Inverse Lithography

ILT was first introduced as a DfM/CAD solution for the 65-nm process node. A 6-T SRAM array bit cell with split double word line (SDWL) architecture, a design lithographically well behaved and manufacturable in large matrix densities (e.g., 72 MB) [39], was used as the test case, with 2D complexity of the cell due to the abutting line ends. Uncalibrated ILT with scalar optics tested against calibrated MB OPC with vector optics on a dark-field pattern [40] showed pattern fidelity comparable in terms of CDs, line ends, and LER across a similar process window, with corner rounding slightly better for the ILT (Figure 3.22). With calibration and an equivalent optical model, ILT can outperform OPC [41].

While OPC adds line end hammerheads and staircase-type edge corrections with sharp corners, the ILT pattern introduces oval contour lines. The natural smoothing of the drawn staircases at mask writing due to ebeam proximity and viscosity of the photoresist can actually benefit the ILT pattern reproduction, also making it possible to use low-cost laser writers instead of

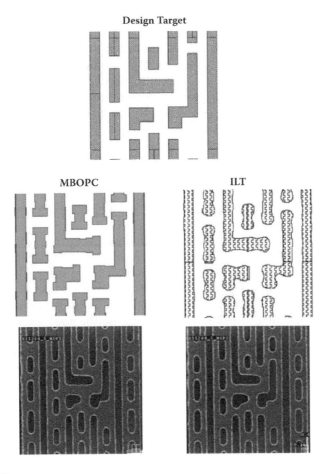

FIGURE 3.22
ILT images of the interconnect layer in a standard cell. (From Balasinski, A., Moore A., Shamma, N., Lin, T., and Yang, H-H. 2005. 25th Bacus Symposium on Photomask Technology. Proceedings of the SPIE. 5992, 881–885. With permission.)

the e-beam tools for a full-fidelity ILT mask. Superior corner rounding for ILT was due to the fact that ILT calculates mask geometries based on the required silicon image, unlike MB OPC, which works from the mask pattern.

3.4.2.4 Mask Manufacturability

An important aspect of 2D DfM and variability reduction is the quality of the information carrier from the virtual to the real space, i.e., the photomask. Pattern transfer from design to wafer traditionally uses a two-phase approach:

- Phase 1: Pattern transfer from design to mask, by direct writing on mask resist with a laser or ebeam tool.
- Phase 2: Pattern transfer from mask to wafer, by exposing the mask image on the photoresist using the light source of the stepping tool.

Therefore, photo masks are a critical component of the ICM flow from the standpoint at both DfM and MfD efforts. Mask quality has a direct impact on fab performance and project schedule, and mask DfM is a part of IC DfM in all its aspects: mask reliability, parametric performance, and yield.

For successful mask DfM, one needs to optimize both Phase 1 (layout) and Phase 2 (mask) of pattern transfer from design to wafer (Figure 3.18). In particular, subresolution defects on masks may interfere with design intent and manufacturing capabilities if not identified by the rule checks. Even for designers working closely with manufacturing, it is difficult to understand and mitigate the impact of mask complexity driving down mask defects and their impact on product performance (Table 3.10) as there is no direct feed-back from fab yield to mask shop.

Mask data preparation (MDP) for design databases can introduce defects due to OPC or fill pattern requirements, such as illegal shapes or off-grid errors (Table 3.11). Mask writing tools interpret these defects by snapping geometries to grid or integrating small features with process-based sizing of the devices. This results in die-to-database (DTDB) mismatches dealt with by ignoring (detuning inspection tools) to release the masks even with non-resolvable data. But, this process of ignoring physical defects on masks by waiving minimum CD rules or allowing for misshaped geometries compro-mises mask fidelity as a medium of design information transfer to wafer. As a remedy, if a mask cannot be confirmed 100% true to its data, the mask shop can be supplied with extra information for an additional verification step to support audit of critical features in poorly inspectable areas.

One should expect that only fully verified data, printable on masks, repro-ducible, and protected by a sufficiently wide process margin, should be trans-ferred to mask shops. However, up to 13% of masks carry design data, which are unresolvable, that is, it does not comply to the basic mask DRC (MRC) checks, such as width and space (i.e., errors have been identified but waived, or the MRC check was not run at all), project schedule requirements and

TABLE 3.10

Impact of Device Properties on Mask Complexity

Design Element	Defined at	Critical Features	First Used at	Mask Complexity Impact	
				Direct	Indirect
Active devices (MOSFETs)	Device library	Active and poly masks, implant OPC, pattern density	Standard cells	Low	Very high
Passive devices	Device library	Pattern density, critical dimensions of precision elements	Block design	Low	Moderate
Interconnects	Manual or automated routing	OPC, pattern density,	Block design	Moderate	Moderate
Fill pattern	Postlayout processing	Pattern density impact of tile size and shape, OPC impact, extractability	Die-level assembly	High	N/A, postprocessing
RET features	Standard cells or postlayout processing	Minimum CD of RET features, context sensitivity, reproducibility	Postlayout	Very high	N/A, postprocessing

Direct impact: at drawn level
Indirect impact: with all assist features

the experimental nature of some structures overrule the clean PV (physical verification) approach.

The signature of mask resolution errors (i.e., missing or extra chrome identified by the inspection tool by comparing to mask data), when compounded with corner rounding, can bridge or pinch the designed lines (Figure 3.24). Eliminating such errors by running MRC CAD algorithms may only be undesired e.g., for testing of the limits of mask writing or inspection tools as part of the research and development [R&D] effort. Therefore, waiving or desensitizing error detection due to the custom R&D verification standards has to be done carefully as MRC errors obscure design intent and make it difficult to understand whether mask inspection can be accepted or rejected.

Defects with the smallest and largest CDs have the lowest impact: the first ones as non-printable, the second ones as occurring infrequently and often detectable by standard design DRC.

By location, mask data errors belong to three main categories (Table 3.12, [42]). Their largest population comes from the scribeline structures and is due to the metrology and process development modules. Whether intentional or overlooked, this group of mask data errors should be reduced in

TABLE 3.11

Mask Data Defects Impact and Correction Methods

		Defect Type	CAD Repair Procedure	Design Intent Preservation	Defect Mechanism	Possible Design Impact
1	A	Too small width	Up-down sizing	Context dependent	Mask photoresist can peel and redeposit	Impacts unrelated features
2	B	Too small space	Up-down sizing	Context dependent	Mask photoresist can bridge	CD control
3	C	Too small internal corner-to-corner space	Use error flag as cutout	No	Poor CD control of adjacent features	Device CD control or implant enclosure
4	D	Too small external corner-to-corner space	Use error flag as cutout	No	Poor CD control of adjacent features	Device CD control or implant enclosure
5	E	Too small OPC width	Size to required CD	Yes	Same as (1) but at design postprocessing	MOSFET drive or leakage current
6	F	Too small OPC space	Size to required CD	Yes	Same as (2) but at design postprocessing	MOSFET drive or leakage current
7	G	Acute angle	Orthogonalize	No	Poor resolution and CD control	Device connectivity
8	S	Singularity	Use error flag as cutout	No	Extreme case of (3) and (4)	Device connectivity
9	O	Off grid	Snap and check DRC	Contextual	Arbitrary snap at write time	MOSFET CD control

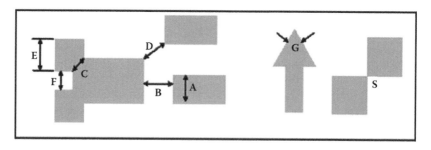

FIGURE 3.23

Types of mask data errors. (From Balasinski, A., Coburn, D., and Buck, P. 2007. Proceedings of the SPIE, vol. 6730, p. 67303J. With permission.)

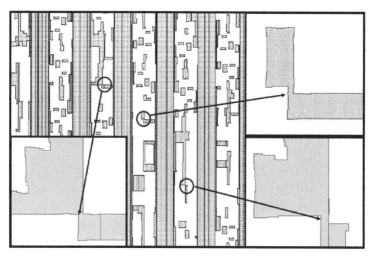

FIGURE 3.24
Examples of mask data errors in a design database that can read to pinching or bridging. (From Balasinski, A., Coburn, D., and Buck. 2007. Proceedings of the SPIE, vol. 6730, p. 67303J. With permission.)

TABLE 3.12

Categories of Mask Data Errors

Location on Mask		Category	Contribution (%)
	1	Scribe/frame	70
	2	Die pattern	16
	3	Lettering, titles, and logos	14
Origin		**Category**	**Controlled by/Issues**
	1	Design intent	Design rules for drawn layers
	2	Dummy pattern (waffling)	Fragments of waffles
	3	OPC/SB/RET	Can merge with each other or with drawn design pattern
	4	Metrology structures	Labels
			Etest/look ahead/corner evaluation
			Metrology structures
			Daggers/leftovers from other technologies
Consequences to mask process	Write time		
	Inspectability		
	Repairability		
Consequences to wafer process	Product parameters (parametric yield)		
	Litho process window		
	Miscalibration for the next technology mode		

the order of the highest risk priority number (RPN)/ROI for the product line (Table 3.13) for masks with:

1. benign but multiple errors
2. moderately severe errors on databases for critical products
3. severe errors for any product

The corresponding approaches to mask data repair may be as follows (Figure 3.25):

1. Flattening the error distribution (trend *a*) using CAD routines verified for preserving design intent. Manual repair of large numbers of errors is not practical.
2. Eliminating the bulk of mask error population (trend *b*) using CAD routines combined with waiving of errors if there are good reasons to do so. The waiver file should be provided to the mask shop for local detuning of inspection tools.
3. Truncating error distribution (trend *c*) by manual edits. CAD routines may first find the root cause of the defects (e.g., type of cell causing the problems).

For low-volume products, considered mask data defects may not be as worth fixing by some designers, if the product schedule and performance are not impacted and a rebuild of high-grade masks would be required to remove nonresolving experimental structures. On the other hand, MRC errors with low risk require priority repairs if they impact volume production or are a result of increased mask complexity due to DfM solutions at the product level (Table 3.14). CAD procedures that eliminate subnominal features in the database (Figure 3.26) by edge smoothing emulate the actual mask-writing process when such features are physically smoothed due to the integrating properties of the ebeam and photoresist.

A postsmoothing review of LVL (layer vs. layer) results vs. the original database may be required to evaluate the impact of MRC correction on design intent.

There are several approaches to quantify the result of mask writing (Table 3.15). A clean result of the preferred mask DTDB inspection is only possible with zero MRC errors. If DTDB fails, the mask shop can either detune the inspection tool or use the less-sensitive die-to-die (DTD) inspection, assuming that a database defect would be reproduced on two (or more) identical dice. However, DTD inspection inspection fails for shuttle masks with no redundant designs (lack of redundancy also limits the ability of the mask shop operator to identify real mask defects, e.g., due to particles).

Any errors identified at DTD inspection would require manual mask disposition. A "blind eye" inspection can bypass the unresolved data by creating "do not inspect regions" (DNIRs). This way, the inspection is expedited with minimal customer involvement.

TABLE 3.13

Categories of Mask Defects and Their Impact on a 10-Point FMEA Scale

| Size | No. | Category | | Impacts | FMEA | | | |
		On Mask	On Wafer		Severity	Occurrence	Detectability	RPN
Smallest to Largest	1	Not printable	Not printable	Write time	2	5	2	20
	2	May peel	Potential repeaters	Inspectability	10	5	8	400
	3	Marginal CD control	Not printable	Write time	4	5	5	100
	4	Poor CD control	May peel	Hot spots	8	4	5	160
	5	In spec	Marginal	Poor CD control	7	3	2	42

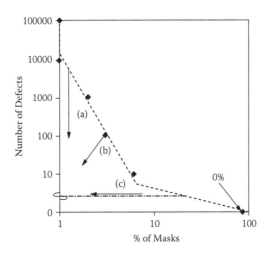

FIGURE 3.25
Mask data error elimination options and trends. (a) multiple benign errors; (b) moderately severe errors; and (c) few severe errors. (From Balasinski, A., Coburn, D., and Buck, P. 2007. Proceedings of the SPIE, vol. 6730, p. 67303J. With permission.)

It is ironic that in the mask shop-to-design feedback, uninspectable masks can be interpreted as a problem of the mask shop, not of design. The IC design house, rather than providing good mask data, may choose a different mask shop to "resolve" the issue being their own fault.

If subresolution mask structures are required, MRC errors need to be reviewed by designers to make sure that an internal DRC run at the mask

TABLE 3.14

Mask Complexity Impact of Product-Related DfM Solutions

Design Issue	What to Improve: Technology Issue	How	Mask Impact
Exponential dependence of leakage on CD (e.g., $\Delta CD = 5$ nm, $10\times$ leakage) incompatible with handheld applications	10% horizontal (CD) variation range too wide for sensitive circuits MEEF needs to be monitored	Define EPE/ORC rules Use high; NA litho tools with optimized illumination	Requires mask CD control down to 1-nm level (MEEF!)
Capacitive coupling for RF as critical as resistive path for DC	Reduce vertical dimension (planarity) range but use sparse fill pattern	New fill or full extraction	Density of fill pattern and spacing checkable at mask level
Mismatch among identical IP blocks (standard cells)	Layout environment includes OPC and density rules	Layout orientation and placement rules	CD control down to 0.1% for nonadjacent features

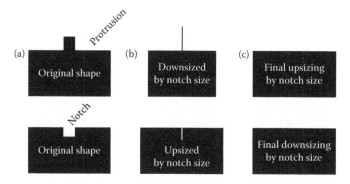

FIGURE 3.26
CAD procedure for mask data error elimination based on upsizing and downsizing.

shop would be adjusted. The mask shop may accept a waiver file to focus the inspection on real mask process errors. If design is not able to repair unresolvable geometries or off-grid points, a finer grid and a mask writing tool with higher resolution should be used.

Mask DfM for future process generations should ensure that mask error checks are routinely performed as the last step of design verification before mask making and eliminate the 10% of errors that are disregarded or not reviewed by design. As a remedy, either experimental structures causing a majority of the problems have to be subject to the MRC standards similar to the ones applicable to design database, with the assistance of tightly controlled waivers, or a higher-resolution layout and mask process have to be used. The alternative (i.e., manual mask inspection and ad hoc interpretation of design intent to override rejects) may create product yield loss; recovery from this would require rebuilding the masks.

TABLE 3.15

Mask Inspection Strategies

No.	Strategy	Pros	Cons
1	Inspect at full sensitivity Manual error categorization	High probability of catching all defects	Risk of miscategorizing Time-consuming manual review of defects
2	Desensitize detection until inspection completed	Low sensitivity to nuisance defects; no need of waivers from customer	Low sensitivity to real defects; mask with defects can be released
3	Iterative detection and "DNIRing" of problem features	Real-time decision by inspection operators	Numerous iterations, inefficient use of expensive tools
4	MRC checks to find and reconcile unresolved data	Decisions oriented at successful completing of inspection	No severity consideration from device or yield perspective

In summary, there are many ways to ensure quality layout with high-resolution OPC and low variability. The intuitive method for layout variability reduction to use geometries several times larger than the minimum CD enabled by technology is not always the optimal solution. While lowering the variability of optical distortions at corners and line ends of layout geometries would call for a more stable, high value of the k_1 factor (i.e., less-advanced OPC) as the rounding resulting from optical patterning does not scale with the technology node but with the wavelength, across-field variations increase with the field size. Optical and mechanical proximity variations can be reduced instead by adding a guard band of dummy geometries up to the screening distance of α times the patterning wavelength (e.g., 4λ). But, when resolving architectural problems of sensitive circuits by increasing the layout footprint, one needs to account for both the increased parasitics, which compromise device cost and applications, and the possible long-range interactions, which would change the yield response (e.g., pattern density).

One should note that on-layout or postlayout DfM corrections may not align with the CBC design flow. They focus on a quick solution instead of addressing the root cause of the hot spots. An approach consistent with CBC and the rule of 10 is to propose optimal layout geometries at the electrical design stage, based on restricted choices from a few categories with properties predefined by CAD, layout, mask shop, and manufacturing.

3.4.3 3D: On and Off Die

Variability reduction in 3D should be considered in two aspects: on die and off die. On die, it is related to the planarity of the IC layer structure, which is prerequisite to enabling tall stacks of connecting layers stitched with contacts or vias with high aspect ratios (ARs; i.e., the ratios of layer isolation thickness to layout CDs). Unlike the 2D variability related to optical wafer patterning, the processes and mechanisms responsible for 3D variability have changed several times throughout the IC technology generations as more advanced planarization techniques were developed, allowing the contact AR to increase (Figure 3.27). For technologies for which the contact AR was around unity, a reflow of interlayer dielectric (ILD) was sufficient to isolate stacked metal layers from each other. As the contact AR increased, multiple deposition and etchback steps were needed to prevent keyhole (air gap) creation. At one point the use of CVD (chemical vapor deposition) instead of PVD (physical vapor deposition) enabled conformal ILD deposition. As the planarity was further improved by CMP, 3D layout rules became necessary to control the parasitics of interconnect routing, such as interlayer shorts, high contact or via resistances, and the parametric quality of RLC elements.

3.4.3.1 Planarization

Etching of a mask pattern printed on the surface of a wafer creates a permanent change in the topography of the etched physical layers. The resulting

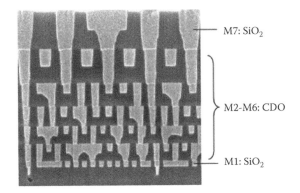

FIGURE 3.27
Increasing aspect ratio of ILD. (From C-H. Jan et al. 2003. 90 nm generation, 300 mm wafer low-KILD/Co Interconnect Technology, *IITC*. With permission.)

pattern density of these layers would vary, having an impact on the vertical CD distribution (thickness) of the ILD deposited over them. The subsequent ILD planarization necessary to even out thickness variations may not prevent parametric variations of the product. With process scaling more pronounced in the x-direction than in the y-direction, the cross-sectional AR of devices and interconnects increases, and to connect or isolate the individual devices, planarity requirements of all the layers in the IC are becoming more stringent [43].

This creates a need to add the fill pattern (waffles, dummies) in order to:

- equalize pattern density for STI etch (ST, active level) to control isolation properties
- control optical and etch proximity for poly gates and metal lines
- control via resistance between metal layers by reducing dishing:
  - for damascene metals, the key parameter is density of metal routing
  - for subtractive metals, the key parameter is density of sized metal routing.

As a result of these requirements, waffling (or dummy fill) routines would differ among the layers (Table 3.16). The waffling parameters have an impact on the following parameters of layout pattern density:

- absolute value
- range
- gradient
- distribution over the die, frame, and reticle area

All these parameters have an impact on both physical and electrical properties of the product die.

TABLE 3.16

Mask Dependent Permanent or Transient Die Pattern and Impact of Pattern Density Variation, ΔPD

Type of Change	Layers	Impact of ΔPD	Comments
Permanent	Active	Poor device isolation: Distribution of punch-through or breakdown voltages	Flat surface required for accurate poly patterning (depth of focus)
	Poly	Poor planarity: • High contact via resistance • Poor control of interlayer capacitive coupling	All contacts etched through the same ILD should clear at the same time to prevent resistance variation due to local overetch
	Metal	• Reduced depth-of-focus during photolithography	Uniform ILD capacitance required for consistent timing for RF products for all parasitics to be extracted;
	Contacts, vias	Dishing	Depth of focus
Transient	Implant layers	Shadowing pattern proximity	

Placement of an IP block in the die may create pattern density gradients, and a global addition of fill features (waffles) is required, typically using one or more of the three main approaches (Table 3.17) developed for different product generations.

There are several critical parameters of the dummy fill and the resulting pattern density distribution it ensures (Table 3.18).

Adding dummy structures should level out pattern density among the individual blocks in the die, but different die regions may require different fill (Figure 3.28). Sparse waffling may adjust pattern density on a local scale. If a large number of die regions have to be exempted from waffling using area id "keep out" CAD layer, a denser pattern may need to be added in the adjacent areas to compensate and increase local parasitic capacitance. Note that while, intuitively, exempting die regions from waffling should mean lower overall pattern density, the only way the long-range pattern density can be equalized is by *adding* not by *removing* features. Therefore, the absence of fill pattern in some areas could trigger a significant increase of extra pattern density everywhere else, even at the expense of short- or medium-range density gradients.

Fill pattern extraction may be based on the cross sections of metal lines without and with the under-, intra-, and overlayer waffles (Figure 3.29). The nonintrusive fill pattern at the IP block level can be based on permanent placement only in timing-critical IP blocks. In the remaining locations, fill placement would be done at the die level, as required by a pattern density check. The parasitics,

TABLE 3.17

Dummy Fill Pattern Methodologies, with Algorithms Depending on the Type of the Etch Process: Subtractive (e.g., Poly) or Damascene/Trench (e.g., Metals)

Approach	Pros	Cons
Manual (drawing features in design layer)	Enables pre-integration adjustments of pattern density and extraction of all parasitics	Requires guidelines how to hand draw fill pattern in the various sections of the die; may require iterations
Geometric (uniform fill pattern added by CAD to all empty areas)	Fixed and known density, simple algorithm	Uniformity may require high pattern density, disadvantageous for RF/analog products (capacitive coupling of sensitive signal paths); CAD iterations increase design cycle time
Intelligent	Initial die pattern density evaluated, first followed by adjustable fill of density depending on the gradient at block level	Expensive tools, long run times, not fully extractable results may differ from run to run depending on the computing engine

extractable more efficiently at optimal block level, would be for the final SPICE simulation, frequency response, and low phase noise as inserting the fill pattern at the die level does not allow for full electrical verification and should be minimized. Because SoC products are made of IP blocks with various pattern densities, one can prefill these blocks within the range recommended by the

TABLE 3.18

Parameters of Dummy Fill

Parameter of Dummy Fill	Comments
Density (absolute)	Should be kept low after fill pattern addition to reduce parasitic capacitance, usually not critical for processing
Density (range)	Critical for uniform distribution of IC parameters; typically in the 20–30% range across die; also impacts macroloading at etch
Density (gradient)	Important for damascene process due to macroloading at etch
Waffle size	Trade-off between easy placement, low capacitive impact (requires small size), and low database impact (requires large size)
Density of sized waffles	Can reach 100% if waffles close to each other in subtractive etch process
Number of added waffles	As low as possible, but if no waffles detected in some areas may confuse during the review that the algorithm is not working correctly
Run time	As short as possible

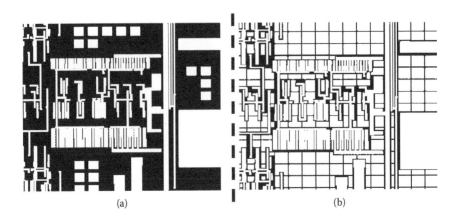

FIGURE 3.28
Two densities of fill pattern: (a) sparse; (b) dense.

fab (e.g., 30%) before placing these blocks in the die. Then, correlating pattern density between the block and the die, based on

block location in the predefined area with average die pattern density

pattern density of the neighboring blocks

placement of power grid or other wide metal busses over the block

should minimize the electrical impact of fill pattern on product parameters.

Placement of dummy features without detailed design rules can be detrimental to device performance [44]. While device matching can be improved, the goal of CAD-based waffling is to accomplish uniform pattern density at different ranges specified by the size of the rule check window (Table 3.19). Fill pattern placement is disturbed by keep-out regions, such as due to the vicinity of devices (short range), substrate or well boundaries (midrange), or large blocks of precision RLC elements (long range) (Figure 3.30). Simulations of pattern density are required to verify its distribution, also depending on the etch process (subtractive or damascene). It is important to experimentally verify the design of passive devices and their governing rules before they are implemented in real products. Precision poly resistors restrict the adjacent geometries and require large areas to meet matching requirements. Process simulations are needed to optimize the fill pattern for low impact to the quality factor (Q), inductance, and self-resonance frequency of a differential inductor in top-level metal on a high-resistivity substrate. Circuits with large high-Q inductors (100 µm × 100 µm) disturb distribution of pattern density but increasing this density by adding waffles would reduce their self-resonance frequency.

Variations of threshold voltage (V_T), resistance, or inductor parameters in RF SoC are already impacted by the circuit location with respect to the edge of the die, I/O pads, or electrostatic discharge (ESD) clamps. The presence of unrelated metal close to an inductor affects its inductance, Q, and the frequency

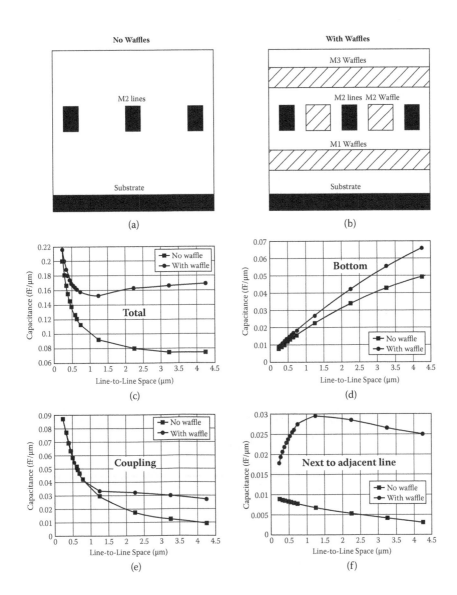

FIGURE 3.29
Fill pattern extraction depending on the position of waffle layer. (a) cross-section without waffles (b) cross-section without waffles; (c)-(f) correlations between capacitance and line-to-line space.

of self-resonance. In applications such as voltage-controlled oscillators (VCOs), which rely on a high-Q inductor (L) and variable capacitor (C) tank for tuned amplifiers, the characteristics of the inductor must not be affected by fill pattern.

Two-dimensional inductors already have poor performance, which can be further aggravated by the presence of other structures on the die. They also present modeling challenges involving multiple materials with distributed magnetic constants. Large inductance values cannot be accomplished

TABLE 3.19

Functions of Fill Pattern at Different Interaction Ranges

Interaction Range	Function of Fill Pattern	Level	Model Deficiencies
Short, 1 μm	Width and space CD uniformity improvement due to optical proximity control; may reduce the aggressiveness of litho correction features (OPC) and widen the lithography process window	Device	Models of microloading included based on litho simulations
Medium, 10 μm	Control ILD composition related to its deposition and etch process	Block	Etch bias substituted by photobias models (black box model)
Long, 100–1,000 μm	Reduces ILD thickness variations after CMP and improves contact resistance uniformity.	Die	Requires planarity simulation at die level

due to the high cost of the die area. At the same time, attempts to successfully emulate inductance behavior in analog circuits by logic gates across a range of parameters have not been satisfactory. Experiments are therefore needed to optimize and customize inductance elements depending on the number of metal layers, thicknesses of the ILD, and product applications.

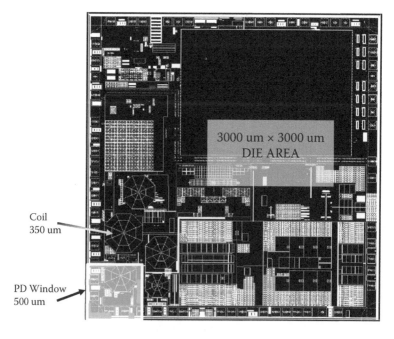

FIGURE 3.30

Inductors in an RF die, impacting the pattern density (PD) distribution.

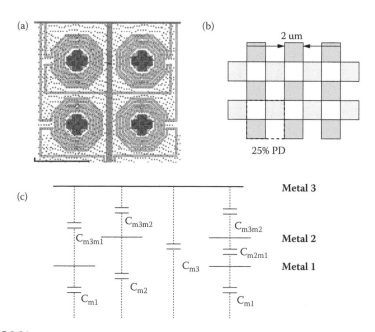

FIGURE 3.31
Fill pattern in inductors. Intra- and interlayer metal fill capacitances: a) example fill pattern coverage over inductors; b) A checkerboard pattern of metal 2 offset by one metal square in the x- and y-directions from the same checkerboard pattern of metal 1 is used in simulation of least-capacitance penalty; and c) serial arrangements of: $C_{m3} < C_{m3m1} C_{m1}/(C_{m3m1} + C_{m1}) < C_{m3m2} C_{m2}/(C_{m3m2} + C_{m2}) < C_{m3m2} C_{m2m1} C_{m1}/(C_{m3m2} + C_{m2m1} + C_{m1})$, because $C_{m3} < C_{m2} < C_{m1} < C_{m3m2} < C_{m3m1} < C_{m2m1}$. (From A. Balasinski and B. Madhaven. 2008. Impact of fill pattern on RF response of passive elements. VLSI-SOC. Conference on Very Large Scale Integration Systems. With permission.)

The fill pattern overlying and underlying the actual inductor (Figure 3.31) [45] may be added in a relatively nonintrusive way as compared to manual metal fill over the entire inductor area, just to provide the minimum required amount of metal coverage. The optimal fill pattern should minimally have an impact on Q, L, and the self-resonance frequency of a differential inductor (Figure 3.32) built in top-level metal on a high-resistivity substrate. Excluding the inductor and its surrounding space from fill algorithms will prevent changes to their characteristics, assuming that the final pattern density would not result in thinning of metal wiring (e.g., at lower-metal underbridges in the inductor layout). In circuits with clustered inductors, large areas excluded from the fill pattern may give rise to highly nonuniform pattern densities across the die leading to degradation or functional failure of the IC due to the thinning of ILD in the areas of low pattern density. On the other hand, pattern density fill disregarding special requirements of the inductor will not only lower its Q and self-resonance frequency due to capacitive loading (electric field effect) but also may reduce its inductance (magnetic field effect).

It is preferred to use large banks of inductors. For example, the LC-VCO may use eight differential inductors that are several hundred μm each, with

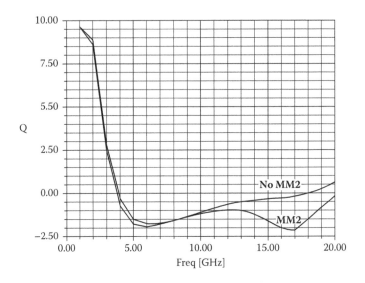

FIGURE 3.32
Frequency-dependent Q factor with and without fill pattern on metal 2. (From A. Balasinski and B. Madhaven. 2008. Impact of fill pattern on RF response of passive elements. VLSI-SOC 2008. Conference on Very Large Scale Integration Systems. With permission.)

characteristics without fill on a 20-GHz network analyzer corresponding to a differential inductor model, with total inductance of 6.2 nH, maximum Q of 13 at 2.2 GHz, and a self-resonance frequency of 6.5 GHz. Ansoft Designer [45] matched the 3D simulations created by importing the inductor layout into Ansoft's High Frequency Structure Simulator (HFSS; [45], Table 3.20). Pattern density fill had an impact on the inductance and self-resonance frequency by the least amount for a checkerboard pattern, avoiding regions directly underneath the inductor traces and filling its central and external region at a spacing equal to the trace width of the inductor (Q reduced from 13 at 2.2 GHz without fill to 9.73 at 1.5 GHz with fill). Measured phase noise and frequency of the LC-VCO with eight inductors with such fill were in agreement with simulations.

Small squares of metal fill pattern created the lowest disturbance of magnetic field characteristics of the inductor. Figure 3.31 shows that to reduce capacitance to substrate, staggering checkerboard patterns of squares of metal in different layers were preferred to stacking them on top of each other. RF characteristics of the inductor were most significantly affected by fill directly underneath the inductor due to capacitive loading. The impact of fill in the central area of the inductor, where the magnetic flux is confined, and of the spacing of fill to the outer and inner edges of the inductor evaluated by 3D electromagnetic simulations, indicated the need to use larger metal squares, 10×10 μm, to reduce the size of the model in the 3D solver (the actual fill pattern waffle size was 2×2 μm). For interior fill simulation, a larger metal

TABLE 3.20

Summary of 3D Simulation Results of Uniform Checkerboard Pattern Density Fill Impact on Self-Resonance Frequency of Inductor

Model	Self Resonance Frequency (GHz)
Measurements	**6.5**
M1 under inductor, M2 outside inductor	3.65
M1 under inductor, M2 outside central region of inductor	3.0
M1 under inductor, no M2	3.85
M1 + M2 in central region only	6.5
M1 + M2 in central region and exterior of inductor spaced from inductor at a spacing equal to the width of the inductor traces	6.4

Note: The simulation took 44 hours on an eight-core Intel Xeon 5355 with 32 GB of RAM for 720,000 tetrahedra at accuracy criterion of 0.02. Increasing the accuracy to 0.01 added 24 hours of CPU time, resulting in a model with 2 million tetrahedra and a RAM footprint of 14.9 GB. The difference in results was less than 2%. The self-resonance frequency of the inductor was 6.4 GHz, versus 6.5 GHz of the original model.

M1, metal 1; M2, metal 2.

square would reduce model complexity to minimize eddy current losses that would lead to a reduction in the inductance. While 3D EM simulation results showed that the self-resonance frequency of the inductor was not changed from 6.5 GHz of the original layout, any fill pattern directly underneath the inductor traces resulted in a greater than 40% reduction in that frequency.

Interior and exterior fills were spaced away from the inductor at an edge-to-edge distance equal to the trace width of the inductor. The maximum value of Q was reduced from the measured value of 13 at 2.2 GHz without fill to 9.73 at 1.5 GHz, which was acceptable for RF applications. Such a pattern density fill strategy was in agreement with simulations of the LC-VCO based on inductor models with no pattern density fill.

Similar to the 3D electrical impact of fill pattern on the distributed parasitics, the physical impact of dummy pattern is also related to its interaction range (Table 3.19) with 3D topology on the wafer. Short-range (a few micrometers) pattern density variations may have an impact not only on optical imaging or planarity but also on chemical composition of the etched layers. This is because the composition of sidewall spacers may depend on the topography of the underlying gate pattern and the ILD etch rate. The resulting etch selectivity may have an impact on electrical isolation between the damascene tungsten filling the self-aligned trench contacts, and the impact underlying poly gates protected by spacers. Dummy placement, based on the interaction range of ILD deposition and etch controls tungsten trench shorts and affects routing capability.

3.4.3.2 Etch Selectivity Control

The features of dummy fill depend on what is required to compensate proximity effects for lithography, etching, and CMP. The kinetics of the etch process depend on the topography of the etched surface [46]. For poly or metal lines over wide, open surfaces, in situ redeposition of the etched ILD material is marginal. Oxygen plasma for ILD etch increases the deposition component from the silane, oxygen, and dopant source. When RF bias is applied to the wafer substrate, oxygen and the inert gas are ionized and accelerate toward the wafer, sputtering deposited material. As a result, the ILD is simultaneously deposited and sputter etched, allowing gaps with high ARs to be first filled, followed by the ILD layer reflow for planarization [47]. For wide spacings between the lines, there is no "breadloafing" and "keyhole" formation (Figure 3.34), and chemical composition of the ILD is uniform. But, a high AR gap fill with high-density plasma (HDP) and CVD gas mixture containing oxygen, silane, a dopant source, and an etching inert or noble gas simultaneously depositing and etching the ILD layer [47] may compromise integration scheme of the self-aligned contact (SAC).

SAC architecture tightens the spacing rule to the adjacent gate. It requires different types of ILD to prevent shorting between the contacts to active area and MOSFET gates. Narrow and deep contacts to active, area adjacent to transistor gates (Figure 3.33) encapsulated in a spacer material as an etch stop, require more overetch cutting through the ILD to clear the openings of

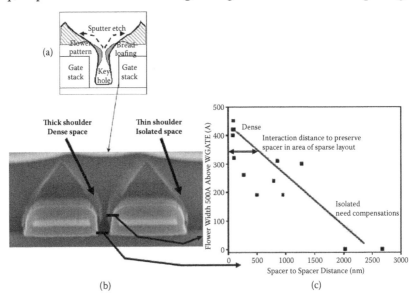

(a)   (b)   (c)

FIGURE 3.33
SAC architecture. (a) mechanism of sputter etch from the keyhole; (b) spacing-dependent shoulder formation; and (c) correlation of space and shoulder width. (From O. Pohland et al. 2007. New Type of Dummy Layout Pattern to Control ILD Rate Proceedings of the SPIE. vol. 6798, p. 679804. With permission.)

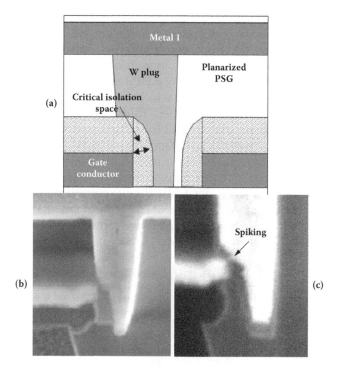

FIGURE 3.34
Keyhole effect. (a) a conceptual cross-section of self-aligned contact showing design and deposited layers; (b) cross-section of the area shown in (a); and (c) contact spiking to gate due to over-reach. (From O. Pohland et al. 2007. New Type of Dummy Layout Pattern to Control ILD Rate Proceedings of the SPIE. vol. 6798, p. 679804. With permission.)

different depths [48]. ILD deposited over field effect transistors (FETs) should be planarized before the contact etch. The self-aligned etch of ILD is selective to the spacer material, to the protective layer, and to the active area. For contact AR over 4:1, complex etch chemistries attack the contact bottom and sidewall spacer material, resulting in current leakage due to the high electric field in the gate corner. Dense topography of the memory array requires longer etch times to reduce the keyholes and reentrant angles. While periphery contacts may have already been cleared, their spacers may not withstand the aggressive overetch even with high selectivity. The sputter etch of ILD-depleted dopant deposits a feature called a "flower pattern" at the top of the gate stacks. Because this flower pattern is dependent on the distance to the gate poly, SAC etch differs between the densely spaced gates and the open area. Longer etch time is required to clear the contacts in the dense area, but it overetches the spacer in the periphery. This shearing of the SAC to the MOSFETs can be prevented by an etch-specific dummy fill to ensure the ILD etch deposition kinetics is balanced between dense and sparse regions at a medium interaction range.

Three MfD and DfM solutions were considered to reduce the flower pattern shearing. First, because the process parameters: HF/RF radio frequency power and ILD dopant concentration were easy to control, the MfD solution of SAC redevelopment was rejected as more time consuming than adjusting the mask pattern. As the second option, a DfM approach disallowing SAC in the periphery was rejected as the one which would result in significant cost of relayout and larger footprint. The best, third option, was layout postprocessing by CAD, eventually allowing for the self-aligned trench contacts to be used in both dense and isolated layout environments (Figure 3.35).

The dummy poly pattern was introduced in the areas of sparse layout to match the array spacing for the semi-isolated poly edges. In contrast to the fill to address pattern density for CMP, the fill pattern for etch proximity reduction was placed before OPC. Due to its medium interaction range, this fill pattern may be needed to improve gate spacer protection and enable local interconnect trenches to cross over them. Custom etch may be required for

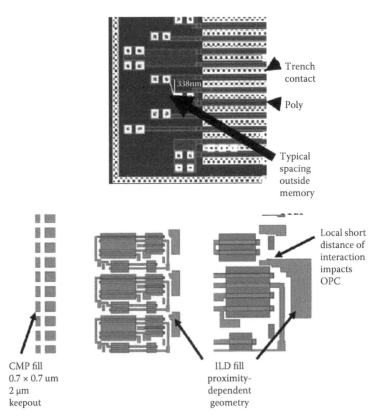

FIGURE 3.35
Layout giving rise to flower deposition and the fill pattern proposed to avoid this effect. (From O. Pohland et al. 2007. New Type of Dummy Layout Pattern to Control ILD Rate Proceedings of the SPIE. vol. 6798, p. 679804. With permission.)

sensing elements in SoC devices (photo sensors, chemistry sensors, etc.), and dedicated DfM algorithms should help achieve the required uniformity of electrical parameters across die and wafer.

3.4.3.3 Off-Die Architectures

For off-die implementation, 3D rules should help define the architecture of system in package (SiP) product. The cost-benefit calculation of integrating a complex circuit into an SoC or an SiP has to consider multiple functionalities, such as

- data (signal) reception and sensing: light, color, movement, pH, humidity, sound, radiation
- data translation (analog-to-digital converter [ADC], SAR)
- data storage (DRAM, SRAM, NVRAM, etc.)
- logical operations on input data versus programmed state (core logic)
- generating output signal (DAC) and sending it to the actuator (RF)
- acting on the output signal in the actuator (power devices)

to drive new market applications.

A decision to chose a SiP or a SoC has to be made early in the design process [49]. As discussed the 80/20 rule stipulates large up-front cost commitment for developing and integrating the many components. The cost can be reduced by conforming to the legacy DfM rule of avoiding homegrown solutions to replace off-the-shelf devices ("do not reinvent the wheel"). Commodity circuits should not be redesigned just to make them fit into the SoC footprint; a multidie SiP may be pursued instead.

Advantages of SoC such as integrated design, manufacturing, and test engineering, while lowering assembly and management costs, create risk to fab yield. Custom SoC components add process steps (physical layers or implants), introducing defects. In addition, RLC elements create pattern density problems. Conversion of SoC to SiP with passive elements on a separate structure may be an economically better solution.

Sensing elements integrated on a chip require a dedicated process flow due to their usually large areas, which would have an impact on layout pattern density (e.g., the pixel matrix in image sensors should have minimal metal density) and compromise process control indices such as c_{pk}. RoI-based choice of product architecture between SiP and SoC may depend on the commonality of process modules for the different layout blocks of the circuits. RoI for SiP and SoC can be compared by

- mask/process commonality dependence on the SoC building blocks
- cost of product 3D integration for the SiP die
- final product quality and market value of both product architectures

For SiP consisting of N separate devices with their respective manufacturing costs of C_i, $i = 1 \ldots N$, to deliver a volume of 1 wafer per device would cost N separate wafers a sum of $\sum_{i=1}^{N} C_i$. By comparison, the cost of an SoC would depend on the cost of the common part $\prod_{j=1}^{M} O_j$ of the wafer manufacturing operations O_j, $j = 1 \ldots M$ (e.g., active, poly, contact, metals) and on the separate processing costs for all custom process modules, such as the implants, additional metal masks, and nonstandard processes required for passive elements (capacitors and inductors) for application-specific circuits.

In SiP products, it is easier to leverage aggressive design rules toward die footprint reduction, while in SoC products, integration of the entire functionality on one die helps avoid costly 3D solutions. Unlike for the SoC devices integrated on silicon due to the common set of design rules for the multiple components, formal verification of SiP by CAD algorithms is trailing behind due to the complex architectural requirements of the new systems. To migrate product DfM from 2D to 3D, the intermediate design and extraction steps need to be automated instead of relying on manual (or visual) verification of the 2D blueprints representing multiangle views of 3D products. Similar to what is being done for the IC design, the blueprints need to be converted into sets of layers, for which verifiable (as opposed to stipulative) design rules would be defined. To ensure the efficiency of different packaging options such as WLB GA (wafer-level ball-grid array), PoP (package on package), and TSV (through-silicon via), incremental improvements considered as transition from 2D to 2.5D to 2.75D and finally, to 3D will be required [50]:

- *2D design*: In the design database, there is no third design dimension. Packaging layers are laid out as parts of 2D design as long as the height (Z dimension of data) of the components does not have an impact on electrical characteristics of the product. In reality, only single-layer printed circuit board (PCB) designs can be done in the flat 2D design domain. Except for packaging layers, die connections require higher design dimensions.

- *2.5D design* representation allows IC designs to use dedicated layers (Z-steps) and their sequence for metal connectivity. PCB multilayer package laminate designs and CAD tools use this format. Extraction of Z-data parasitics can be stored in the technology file.

- *2.75D design* is needed for wire-bond structures in RF connections. This format provides a workable compromise if a detailed 3D information is needed only for selected design elements, such as for wire shapes, and other structures are defined in 2.5D. This reduces memory requirements and calculation times.

- Full *3D designs* is not supported by the existing CAD tools. While active IC components (e.g., transistors) had always been locally modeled in full 3D to account for implantation profiles, next-level circuitry is represented in 2(.5)D abstraction only.

All 3D SoC designs would have to be validated within multi-technology engineering environments. The big step is the change from sequential design processes to highly parallel/concurrent engineering tasks in multiple dimensions. It would require comprehensive models for

- Electromagnetic field effects (coupling, signal, and power/ground integrity)
- Thermal effects
- Mechanical and material effects (M-stress)
- Design of density- or performance-sensitive wire-bond structures
- 3D packaging with 3D design rules

Another aspect of 3D DfM is aligning between electrical (E) and mechanical (M) design. Historically, electrical design was isolated from package design. Lack of overlap in the engineering skills, methodologies, tools, and data management did not stand in the way of product development. The output of E-design is a fully verified GDS (graphic database system) file used for mask production. The M-designers use drawings that include text information with no CAD verification but provide stipulative design rules for connectivity, which bridge the gap between electrical and mechanical designs. The electrical behavior of high-density 3D systems is influenced not only by the connecting metal layers or wire bonds (dealt with by E-CAD), but also by the properties of materials like glue, isolator films, interposers, shields (subject to M-CAD).

From a mechanical viewpoint, the RDL (routing distribution layer) (and other WLB layers) and laminate/lead frame packages manage the mechanical/chemical and thermomechanical stress. As interconnects between the top and bottom package rely on standard layouts, the electromagnetic field simulations have to be done in 3D. At the same time, TSV in conjunction with silicon carrier (SC) only requires 2.5D WLB design flow for basic integration. The real 3D design challenges will come with the introduction of passive component models within the SC for full standardization of PoP.

Major benefits of TSV and SC include

- High vertical integration density (ca. 2,500/mm^2)
- Flexibility due to universal connectivity matrix
- High thermal conductivity, excellent mechanical properties and electrical (RF) performance
- Easy (low-cost) and compact passive integration

To summarize, 3D DfM has been making progress through several phases of the ME design integration road map:

- Phase 1: No ME codesign; chip plugged into a predefined package.
- Phase 2: Chip and package align at the start of design flow.

- Phase 3: Chip-package codesign and tooling based on common flows.
- Phase 4: Chip-package codesign: roll out of EDA tools based on scripting and functional requirements.
- Phase 5: Design flow starts at PCB level to address PCB constraints.
- Phase 6: Mechanical constraints, require 3D models M-CAD, and E-CAD start to merge.

The last two phases have to be driven by the market to provide RoI for EDA tool integration. Because E–M design environments that work at only one computer (point tools) would not ensure integrated data verification, EDA has to supply standardized, file-based design flow for 3D data formats at significant cost to the toolmaker, implementer, and user.

3.4.4 4D: Time and Voltage Domains

Time- or voltage-dependent IC parameters reflect parametric reliability of the component devices from transistor to package level. For analog RF circuits, drift of MOSFET V_T, g_m, and S may skew device performance, even if similar wearout would not cause failures in logical circuits. Because IC performance and reliability are both impacted by the voltage range, one can trade operating conditions and output parameters for short- and long-term functionality. Sometimes, the signal amplitude processed by the circuits is small, one can reduce the power supply voltage range to boost reliability. The trade-off 4D rules (with the fourth dimension being e.g., time, voltage, or frequency) should determine the suitability of a product to work in a given environment. For low (subthreshold) gate voltages, for which the current is an exponential function of V_G, a very small voltage margin (e.g., 0.1 V) may suffice to distinguish the logical levels of 0 and 1. Then, to rely on scalability of voltage levels with technology generations would require operating transistors in the subthreshold region. The proponents of such an approach would have to ensure extremely tight process control, as the exponential impact of process parameters on the output current leaves a very small margin for variability. Enhancing the impact of process variations by the exponential, rather than by the linear, factor can be called "anti-DfM" as it assumes that the process is modeled and executed with unrealistic accuracy.

Assuming the frequency to be the "fourth D," digital devices with clock frequencies in the gigahertz range, driven "rail to rail" and enjoying noise margins of over 100 mV, may not require DRCs more complex than standard 2D W/S/E DRC. But, the PLLs (phase-locked loops), sensitive to phase noise, or operational amplifiers with input signals in the microvolt level require simulation of all layout parasitics and developing rule decks, including orientation, pitch, symmetry, and jog rules driving extraction quality in 3D. Circuits extremely sensitive to variations of MOSFET CDs, drive currents, or contact/via resistances, well within the range of standard process specifications, may show context-dependent variations of frequency response by multiple times.

A differential pair functioning in a digital SRAM cell may fail in the analog amplifier due to the different parasitic requirements.

Variability of design elements not only increases the cost of pattern printability through extra mask, litho, and device engineering effort, but also complicates layout placement rules, which increase die footprint and product development time. For example, reduction of mask complexity or data file size should be achievable through the definition of design primitives. But, any randomness in layer-to-layer offsets, permitted by the W/S/E design rules, may give rise to device-to-device mismatches. To control the variability, design rules would need to be driven by models of short, medium, and long range interactions (e.g., related to macroloading or planarity). Standard model accuracy for active and passive devices (MOSFETs, diodes, and RLC elements) would only assume average material properties and nominal CDs. As the desired level of accuracy for RF applications increases, parameter variations and parasitic effects at the block level need to be taken into account, in the extra dimension of voltage, clock frequency, or operating time (device age), depending on the product application and requiring a full 4D rule deck.

By enforcing standardized layout as the mainstream 4D DfM solution, one would reduce not only product development time but also the bandwidth for product customization. Adding rules for matching of the parasitics is difficult due to the complex layout interactions in electrical and time domains. Adding a rule to reduce device variability in 4D can be done as follows:

- annotate critical devices (e.g., MOSFETs) with id layer(s)
- update contextual design rules for these devices (proximity, orientation, guard rings)
- tape out data and build masks
- transfer annotation or id layer(s) to the fab to indicate CD metrology locations of sensitive devices

The RoI of the 4D low-variability design rules would depend on their cost and time of implementation versus cycle time reduction due to

- shorter product development time
- faster production ramp-up
- improved time-dependent manufacturability

Yield and performance improvement due to the low-variability rules may require cycles of learning (CoLs) on silicon to ensure these rules are adequate for real products and low-variability pCells are integrated into key circuit elements first characterized on test vehicles. According to one solution, die architecture consisting of standard cells should be laid out on a grid in a fixed orientation to ensure consistent process bias for all devices. The layout parameterization should include 3D parasitics of on- and off-die wiring, vias,

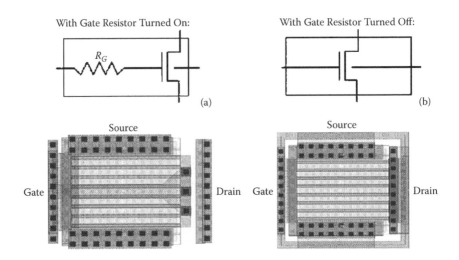

FIGURE 3.36
Two options for gate connection in RF MOSFET (a) single-contacted; (b) double-contacted.

and contacts. MOSFET frequency response can be impacted, for example, by the resistance difference between the single- and double-contacted gate (Figure 3.36), requiring an area increase by 20%.

A process-related example of 4D DfM are antenna rules, which prevent gate oxide from plasma-induced damage related to interactions of wafer topography with plasma in the processing chamber [51]. Contacts or interconnects attached to MOSFET gates (or other areas sensitive to in-process charging) collect ions from the plasma during either the etching (contacts, vias) or overetching phase (subtractive conductors) of the pattern definition process. Antenna rules define the ratio of the antenna area to the MOSFET gate to which it is attached. While protecting the gates can be done by reverse-biased diodes, care must be taken to avoid junction leakage. Deriving DfM guidelines related to antenna rules requires detailed knowledge of process physics (Table 3.21), controlled by the geometry of the plasma chamber, wafer topology, etch time and bias, and so on.

3.5 Summary: DfM vs. Process Variability

The key conclusion from the discussion in this chapter is that there is no universal methodology to pursue device shrinking combined with variability reduction in multiple dimensions. Mask pattern alteration by CAD algorithms can improve local manufacturability at low CD but degrade consistency of layout performance. Improving CD printability in 2D by aggressive OPC reducing corner rounding or line end pullback (low k_1) may create line width ringing problems (i.e., poor image stability), which in sensitive areas lead to

TABLE 3.21

Example of 4D DfM: Antenna Rule Analysis

Plasma-Charging Type	Charging Mechanism	Critical Parameter	Related Process Steps	Antenna Structures	Process Improvement
Equipment related plasma nonuniformity	FN current	Reactor design-induced electrical-magnetic fields, current non-uniformity	Resist ashing, etch near endpoint, plasma strip, sputter deposition	Plates with different areas	Lower plasma density, magnetic field near endpoint
Electron shadow	Angular distributions of ions increase interline charging	Wafer topography, metal, line spacing aspect ratio	Poly/metal/via etch, trench fill, plasma strip	Dense fingers with various spacing and aspect ratios	Reduce aspect ratio, mask layer thickness, overetch, add conductive liner
Microloading	Lower etch rate in routed areas, longer etch with limited chemical supply	Nonuniform routing density, spacing between metal lines	Poly/metal etch, plasma strip	Metal line array with various spacings	Add dummy metal lines routing to for uniform etch
RIE lag	Faster etch, more plasma damage between sparse lines	Nonuniform routing causes aspect ratio effects	Poly/metal etch, plasma strip	Metal line array with various spacings	Transient effect; make etch uniform
RIE ILD overetching	Overetch	Location of metals with vias	Via etch	Via matrix protected by large area plates	Reduce area of lower metal-holding vias, connect large plates to gate
Plasma ILD deposition	Deposition	Deposition temperature, plasma power, ILD thickness	ILD, HD plasma deposition for PSG, TiN sputter	Plates with different thicknesses	Reduce deposition time, temperature, plasma power
Plasma ashing	Stripping	Plasma power, PR thickness, stripping time	Stripping	Plates with different thicknesses	Reduce PR thickness, plasma power, ashing temperature

parametric degradation. Reducing OPC aggressiveness (increasing k_1 at the expense of layout footprint) may increase parasitic RC components and make it more difficult to match the devices over a larger area. Adding dummy fill would improve the nonuniform pattern density in 3D but may increase contextual dependence of the added 3D parasitics. Standardizing the parasitics using large-area pCells would disturb medium-range layout-process interactions and create pattern density matching problems, but randomizing layout distribution over the die area would make it hard to route the signal lines among the blocks or control the IR drop parasitics. Simulating the individual blocks separately for 4D gives no guarantee of their contextual performance on die, but simulating them at die level could require edits to standard cells after their layout is frozen. But despite all these controversies, advanced layout methodology has multiple tools and knobs to deliver the desired solution in all dimensions. As examples, sets of layout CBC best practices are being proposed and supplemented with each new technology generation (Table 3.22). To reconcile the variability conflicts, multiple CoLs may be necessary, delaying the time to market. The economically best solution is to ensure a quick convergence of the DfM approach under the marketing guidance,

TABLE 3.22

Failure Prevention with CBC Layout

Issue	Layout CBC Approach
Connectivity	Full connectivity for all hierarchy levels No extraction errors
Power grid quality	2D grid in all blocks with subsequent metals perpendicular to each other Redundant lines in top 4 metals Pads connected to diffusion away from injection sensitive areas (diffusion, high-Z resistors)
Sensitive regions	Analog, high-Z, sensitive regions must have shielding No parallel lines longer than 5 μm or 2% of length
Matching	Use dummy-ABBA-dummy FETs, extracted LOD, orientation, gate extensions, RLC Same orientation of source drain Dummy structures tied to the same node
DfT	Use LVSable probe points connected to top metal at critical nodes
Wells	Reduce well areas to reduce impact on die footprint
Line W Contact count	Line W > 2× min CD for wide devices, 4× for ultrawide, use 8 contacts or vias for wide buses of injecting node
Latchup High voltage	Substrate contact within < 10 μm Non-Vcc substrate contact at 30 squares around nwell structure Use double-ring and grounded fill pattern around high-risk areas

which would also help categorize the potential risks and failure modes and the acceptable compromise levels from the standpoint of product rating and time to market.

References

1. Aschliman, L. PDK functionality tackles length of diffusion effects. *EETimes*, June 12, 2006.
2. Boone, R., Loparco, D., Melchiori, F., and Thompson, M. *Critical Feature and Improvability Analysis: An Effective Path to DFM Closure*. Mentor Graphics: Willsonville, OR. Mentor White Paper 43893, 2008.
3. Sze, S. M., and Ng, K. K. *Physics of Semiconductor Devices*. 3rd edition. New York: Wiley, 2007.
4. Kahng, A. B., Sharma, P., and Topaloglu, R. O. Exploiting STI stress for performance. *Proceedings of IEEE/ACM International Conference on Computer-Aided Design*, 83–90, 2007.
5. Choi, Y.-K., Ha, D., King, T.-J., and Bokor, J. Investigation of gate-induced drain leakage (GIDL) current in thin body devices: single-gate ultra-thin body, symmetrical double-gate, and asymmetrical double-gate MOSFETs. *Japan Journal of Applied Physics*, 42, 2073–2076, 2003.
6. Crupi, F., Degraeve, R., Kerber, A., Kwak, D. H., and Groeseneken, G. Correlation between stress-induced leakage current (SILC) and the HfO_2 bulk trap density in a SiO_2/HfO_2 stack. *Proceedings of the 42nd International Reliability Physics Symposium*, 181–187, April 2004.
7. Schuegraf, K. F., King, C. C., and Hu, C. Impact of polysilicon depletion in thin oxide MOS technology. *International Symposium on VLSI TSA*, 86–90, 1993.
8. Takii, E., Eto, T., Kurobe, K.-I., and Shibahara, K. Ultra-shallow junction formation by green-laser annealing with light absorber. *Japan Journal of Applied Physics*, 44(24), L756–L759, 2005.
9. Krishnan, A. T., Cano, F., Chancellor, C., Reddy, V., Qi, Z., Jain, P., Carulli, J., Masin, J., Zuhoski, S., Krishnan, S., and Ondrusek, J. Product drift from NBTI: guardbanding, circuit and statistical effects. *Proceedings of IEDM*, 78–81, December 2010.
10. Kim, J. J., and Roy, K. Double gate-MOSFET subthreshold circuit for ultralow power applications. *IEEE Transactions on Electron Devices*, 51(9), 1468–1474, 2004.
11. Yeh, C.-C., et al. AS low operating power fin FET transistor module featuring scaled gate stack and strain engineering for 32/28 nm SoC technology. *Proceedings of IEDM*, 772–775. December 2010.
12. ITRS Roadmap. www.itrs.net.
13. Gupta, P., Kahng, A. B., Muddu, S. V., and Nakagawa, S. Modeling edge placement error distribution in standard cell library. Design and process integration for microelectronic manufacturing IV. *Proceedings of the SPIE*, 6156, 265–276, 2006.
14. Poonawala, A., and Milanfar, P. A pixel-based regularization approach to inverse lithography. *Microelectronic Engineering*, 84, 2837–2852, 2007.

15. Pomplun, J., Zschiedrich, L., Burger, S., Schmidt, F., Tyminski, J., Flagello, D., and Toshiharu, N. Reduced basis method for source mask optimization. *Proceedings of the SPIE*, 7823, 78230E, 2010.

16. Bin, L. Y. New DFM methods enable early yield prediction. *Electronic Engineering, Times-Asia*, September, 2006.

17. Kawasaki, H., Basker, V. S., Yamashita, T., Lin, C.-H., Zhu, Y., Faltermeier, J., Schmitz, S., Cummings, J., Kanakasabapathy, S., Adhikari, H., Jagannathan, H., Kumar, A., Maitra, K., Wang, J., Yeh, C.-C., Wang, C., Khater, M., Guillorn, M., Fuller, N., Chang, J., Chang, L., Muralidhar, R., Yagishita, A., Miller, R., Ouyang, Q., Zhang, Y., Paruchuri, V. K., Bu, H., Doris, B., Takayanagi, M., Haensch, W., McHerron, D., O'Neill, J., and Ishimaru, K. Challenges and solutions of FinFET integration in an SRAM cell and a logic circuit for 22 nm node and beyond. *Proceedings of IEDM*, 1–4, December, 2009.

18. Balasinski, A., Gangala, H., Axelrad, V., and Boksha, V. A novel approach to simulate the effect of optical proximity on MOSFET parametric yield. *Proceedings of IEDM*, 913–916, December, 1999.

19. Si2. MEEF. Mask error enhancement factor. *DFM Dictionary*. June 26, 2008. http://www.si2.org/openeda.si2.org/dfmcdictionary/index.php/MEEF.

20. Yu Ya, C., Wu, Y.-H., Shih, C.-L., Jengping, L., Kan, F., and Lin, J. Effects of mask bias on the mask error enhancement factor (MEEF) for low kl lithography process, photomask and next-generation lithography mask technology XII. *Proceedings of the SPIE*, 5853, 757–766, 2005.

21. Clein, D. *CMOS IC Layout: Concepts, Methodologies, and Tools*. Vol. 1. New York: Elsevier, 2000.

22. Balasinski, A., Karklin, L., and Axelrad, V. An integrated simulation scheme to ensure design shrinkability for sub-100 nanometer technologies. *Proceedings of MIXDES 2001*, Zakopane, Poland, June, 2001.

23. Scott, M. M., Lars, W. L., Azalia, K., and Ioana, G. U.S. patent application 20070261013, 2007.

24. Ishida, M., et al. A novel 6T-SRAM cell technology designed with rectangular patterns scalable beyond 0.18 μm generation and desirable for ultra high speed operation. *Proceedings of IEDM*, 201–204, December, 1999.

25. Verghese, N., Rouse, R., and Hurat, P. Predictive models and CAD methodology for pattern dependent variability. *ASPDAC*, 213–218, January 21–24, 2008.

26. Sylvestera, D., Agarwalb, K., and Shaha, S. Variability in nanometer CMOS: impact, analysis, and minimization. *Integration, the VLSI Journal*, 41, 319–339, 2008.

27. Balasinski, A., Pikus, F., and Bielawski, J. Yield optimization with model based DFM. *IEEE, Advanced Semiconductor Manufacturing Conference, May* 2008, 216–220.

28. Nishi, Y., and Doering, R. *Handbook of Semiconductor Manufacturing Technology*. Boca Raton, FL: CRC Press, 2000.

29. Levenson, M. D. Wavefront engineering for photolithography. *Physics Today*, 46(7), 28, 1993.

30. Schellenberg, F. M., Toublan, O., Capodieci, L., and Socha, B. Adoption of OPC and the impact on design and layout. *Proceedings of DAC*, June 18–22, Las Vegas, NV., 2001.

31. Wong, B. P., Mittal, A., Starr, G. W., Zach, F., Moroz, V., and Kahng, A. *Nano-CMOS Design for Manufacturability, Robust Circuit and Physical Design for Sub-65 nm Technology Nodes*. Hoboken, NJ: Wiley-Interscience, 2008.

32. Kundu, S., Sreedhar, A., and Sanyal, A. Forbidden pitches in sub-wavelength lithography and their implications on design. *Journal of Computer-Aided Materials Design*, 14(1), 79–89, 2007.

33. Gupta, P., Heng, F.-L., and Lavin, M. Merits of cellwise model-based OPC, design and process integration for microelectronic manufacturing II. *Proceedings of the SPIE*, 5379, 182–189, 2004.

34. Balasinski, A. Layout techniques and rules to reduce process-related variability. *Journal of Micro/Nanolithography, MEMS, and MOEMS*, 6(03), 2007, pp 031009-1–031009-8.

35. Agrawal, A. B., Blaauw, D., Zolotov, V., and Vrudhula, S. Statistical timing analysis using bounds and selective enumeration. *Proceedings of Design Automation Conference*, 348–353, June 2–6, 2003.

36. Yang, J., Capodieci, L., and Sylvester, D. Advanced timing analysis based on post OPC extraction of critical dimensions. *Proceedings of DAC*, 359–364, June 13–17, 2005.

37. Postnikove, S., and Hector, S. ITRS CD Error Budgets: Proposed Simulation Study Methodology. May 2003.

38. Orshansky, M., Milor, L., Chen, P., Keutzer K., and Hu, C. Impact of spatial intra-chip gate length variability on the performance of high-speed digital circuits. *IEEE Transactions on Computer Aided Design of Integrated Circuits and Systems*, vol. 22(5), 544–553, May 2002.

39. Balasinski, A., Moore, A., Shamma, N., Lin, T., and Yang, H.-H. Inverse lithography technology: verification of SRAM cell pattern, 25th BACUS Symposium on Photomask Technology. *Proceedings of the SPIE*, 5992, 881–885, 2005.

40. Poonawala, A., and Milanfar, P. A pixel-based regularization approach to inverse lithography. *Microelectronic Engineering*, 84, 2837–2852, 2007.

41. Jia, N., and Lam, E. Y. Machine learning for inverse lithography: using stochastic gradient descent for robust photomask synthesis. *Journal of Optics*, 12(4), doi: doi:10.1088/2040-8978/12/4/045601, 2010.

42. Balasinski, A., Coburn, D., and Buck, P. Mask manufacturability improvement by MRC, photomask technology. *Proceedings of the SPIE*, 6730, 67303J, 2007.

43. Sinha, S., Luo, J., and Chiang, C. Model based layout pattern dependent metal filling algorithm for improved chip surface uniformity in the copper process. *Design Automation Conference*, 1–6, June 4–8, 2007.

44. Stine, B. E., Boning, D. S., Chung, J. E., Camilletti, L., Kruppa, F., Equi, E. R., Loh, W., Prasad, S., Muthukrishnan, M., Towery, D., Berman, M., and Kapoor, A. The physical and electrical effects of metal-fill patterning practices for oxide chemical-mechanical polishing processes. *IEEE Transactions on Electron Devices*, 45(3), 665–679, 1998.

45. Balasinski, A., and Madhavan, B. Impact of fill pattern on RF response of passive elements. VLSI-SOC 2008 Conference on Very Large Scale Integration System-on Chip, Rhodes, Greece, October 2008.

46. Pohland, O., Spieker, J., Huang, C. T., Govindaswamy, S., and Balasinski, A. P. New type of dummy layout pattern to control the ILD etch rate. *Microelectronics, MEMS, and Nanotechnology*, Canberra, Australia, December 2007.

47. Van Cleemput, P. A. In-situ flowing BPSG Gap fill process using HDP. U.S. Patent application 20020052119, 2002.

48. Jost, M. E., and Hill, C. W. Chemical vapor deposition methods and methods of etching a contact opening over a node location on a semiconductor substrate. U.S. patent application 20020123221, 2002.

49. Krenik, W., Buss, D., and Rickert, P. Cellular handset integration—SIP vs. SOC. *Custom Integrated Circuits Conference, Proceedings of the IEEE, 3,* 63–70, 2004.
50. Reisinger, J. 3D design challenges - from WLB to PoP. VLSI-SOC Conference on Very Large Scale Integration System-on Chip, Rhodes, Greece, October 2008.
51. Gabriel, C. T., and de Muizon, E. Quantifying a simple antenna design rule. 5th International Symposium on Plasma Process-Induced Damage, 2000.

4

Fab Implementation: MfD Response

4.1 Introduction

Design for manufacturability (DfM) efforts conclude at tapeout, which is the latest phase for changing the design database into the product database. Implementation of mask data in IC manufacturing has now to create an optimal response to DfM and synergy with the design. It should either take advantage of DfM efforts built into mask layers or create an alternate path for them to become less complex and costly. This chapter discusses the handshake between the various manufacturing techniques and requirements for design and computed-aided design (CAD) to best meet product line expectations concerning the final product features by an effort complementary to DfM, that is, by manufacturability for design (MfD).

One of the key goals of DfM is preservation of shapes and their dimensions in the process of pattern transfer from layout to silicon to guarantee model accuracy for devices and interconnects. Preservation of dimensions in the horizontal (layout or top view) and vertical (cross-sectional) view Table 1.4 in three-dimensional (3D) extraction ensures control of the RLC parasitics. At tapeout followed by the mask-building (i.e., when the design concept is converted to reality), the product architecture has already been decided, including all of its correct by construction (CBC) components. Manufacturing needs now to supplement DfM with MfD efforts by pushing tool capabilities to ensure tight of control critical dimension (CD) in all directions. Numerous contending approaches of pattern transfer technology would impact pattern definition, which decides about the integrated circuit (IC) performance in different ways. The different enhancements to optical lithography, such as mask splitting, extreme ultraviolet lithography (EUVL or EUV), nanoimprint lithography (NIL), and so on [1–3], each calls for different electronic design automation (EDA)/DfM support. Conventional lithography relies heavily on pattern adjustments, while its MfD options are supposed to streamline CAD implementation (Table 4.1).

One of the key measures of success of the subsequent technology generations, largely driven by the equipment updates, is the capability to sustain the

TABLE 4.1

Examples of How to Define a Preferred Approach Among DfM, CAD, and MfD
Based on Response and Cost

	Issue		Response	Cost	Preference	Why Preferred
1	Too high leakage for handheld applications	DfM	Use MOSFETs with longer channels (add design rules for pitch and orientation)	Slower parts IP invalidation	25	Simple substitution
		CAD	New OPC and fill to reduce 3D CD variations	Complex CAD algorithms increase execution time	25	Does not need design change
		MfD	New etch for better line profile	New equipment	50	Common process for all products
2	Device mismatch causing offset of op-amp gain	DfM	Add pitch and orientation rules or cells	Model calibration Higher mask grades	15	New rules may help future products
		CAD	Add dummy devices	Increase block footprint	15	Isolating the problem
		MfD	Optimize exposure to tighten CD budget on wafer (multiple CD targets)	Long process development time	70	Optimal use of equipment
3	AC leakage due to capacitive RF coupling	DfM	Lower capacitance coupling by relaxed layout	Increase block footprint	15	Isolating the problem
		CAD	Reduce fill pattern density	New waffling tool	60	Predictable results
		MfD	New type of CMP slurry or reverse mask	More material use or more masking steps	25	Opportunity for advanced product use
4	Reduced frequency response due to high poly contact resistance	DfM	Increase device dimensions (e.g., to reduce capacitive coupling with dummy fill)	Increase block footprint	15	Improved litho process window
		CAD	Contact doubling and poly widening	New CAD tool or algorithm	15	Does not need design change
		MfD	New contact materials New CMP slurry	Cost of slurry Time to etch	70	Common process for products

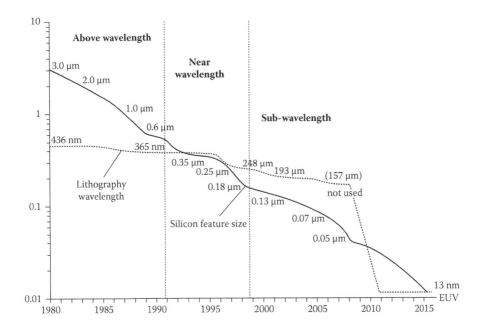

FIGURE 4.1
The subwavelength gap: Slow progress reduction of photolithography wavelength λ versus fast reduction of printed layout CDs.

ongoing CD reductions as a foundation of projected profitability. Doubling transistor density about every 1.5 years from one technology generation to the next by applying a linear scale factor of about 0.70 (Figure 4.1) to all design rules for a product family was the path of lowest cost for design and process, even if done by undermining, crossing over, and eventually grossly exceeding the perceived lithography safety margin, which was due to the advantage of the printing wavelength over the CD of the printed feature. Optimally, one would like to keep legacy design cost at zero by performing such scaling of the mask pattern readily at the mask shop, either without engaging design resources at all or by simple layout touch-ups depending on the shrinkability of

1. Device- and material-related rules, such as minimum isolation, breakdown, leakage, punch-through, and so on, enabled, for example, by pre-optimized implant distributions, material constants, and voltage ranges. In other words, the architectural question, Would it work if we build it? would have to be resolved first at the device level for the direct shrink designs.

2. Minimum CD rules, depending on the shrinkability of pattern transfer methodology, linked to the photoresist pattern, and stepper and mask resolution. This is equivalent to addressing the technical question, Can we build it?

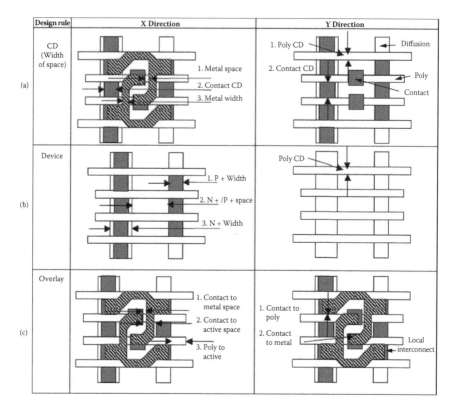

FIGURE 4.2
Key design rules defining the size of a multitransistor memory cell in *x* and *y* directions:
(a) rules related to critical dimensions (CDs); (b) rules related to device functionality at power
supply voltage (e.g., impacting breakdown, punch-through, contact resistance, FET [field
effect transistor] leakage or drive current); and (c) rules related to overlay (e.g., poly to active,
contact to metal).

3. Layer-to-layer overlay tolerance rules (OL) driven by the mechanical
properties of steppers and masks. The tighter tolerances usually
require a more advanced tool set to ensure high values of production
indices (yield, c_{pk}), which brings up the business question, Can we
build it at a profit? The scalability of the static random access memory
(SRAM) cell is driven by a variety of rules: device, CD, and overlay
(Figure 4.2).

Optimizing DfM guidelines and DfM-MfD handshake for the highest
product line profit motivates many companies to look outside the mainstream silicon technologies, with various degrees of success. There are multiple options to build ultradense memory circuits, designed in planar (2D)
or spatial (3D) technologies [4], with a variety of materials. But, in further

considerations, we focus on the dominating silicon planar process as the one with the longest successful legacy, highest DfM leverage, and lowest risk. Given that, the first DfM issue to consider is related to the pattern definition using the different options of the lithography process.

Variability reduction by design (DfM), automation (CAD), and technology (MfD) depends on the trade-off between cost and risk. While setting up a low-variability design flow may delay time to market (TTM), providing a robust product is usually superior compared to derating or all-mask rebuild of an unreliable one (see rule of 10). The return on investment (RoI) of variability reduction is high when process capability is improved at low impact to product design. Therefore, the choices among CBC parametric layout, larger device footprint, extra design rules, or postlayout corrections to mask pattern as responses to the localized issues in a random layout environment (hot spots) need to be compared for minimum intrusiveness.

The option often used to address product DfM issues is to apply automated (CAD) fixes as late in the design process as possible, usually based on the feedback from fab. This path would ensure homogeneity of layout enhancements on a global and local scale (i.e., regardless of the process interaction range). But, bypassing or arbitrarily amending the CBC approach by high-level CAD solutions may negatively project on model accuracy. In addition, it would not allow taking advantage of another opportunity with potentially high RoI, which is the MfD. In Table 4.1, we compare how three approaches—DfM, CAD, and MfD—would help resolve an example issue of leakage reduction for handheld products [5]:

- A design-based DfM solution (i.e., CBC) would call for employing metal oxide semiconductor field effect transistors (MOSFETs) with longer channels, while ensuring that product performance (speed) or die area would not be compromised. However, even if as much as 50% of the short-channel MOSFET population is replaced with their longer-channel counterparts, leakage may be reduced, e.g., by a factor of 2 thereby decreasing the die standby current only by 25%. This may not suffice for the product to be upgraded for handheld applications.

- CAD-based DfM solution would upsize channel lengths of high-leakage MOSFETs without design control. As different blocks may show different responses to the sizing, extensive silicon verification would be required, increasing development time.

- A process-based approach (i.e., MfD) may involve, a new lithography tool or plasma etcher to improve the vertical gate profile and tighten the CD distribution. While burdened with the highest upfront cost, MfD can often be considered the only "real" variability control, valid for all designs. Because the cost and time of redesign (typically multiple man-weeks) impacts the cost of TTM more significantly than the cost of an "in situ" process solution, MfD is often preferred over DfM (viewed as a stopgap approach rather than the root cause fix).

Because fast production ramp-up would mitigate the cost impact of the new tool, the advantage of MfD over DfM can be significant. One can propose RoI analysis (Table 4.1) whereby the preference ratio of MfD to CAD to DfM can be as high as 50:25:25 depending on the estimated effort, resources, cost of time, and hardware. CAD-based DfM would become more attractive if MfD adds incremental processing cost per wafer (e.g., by requiring additional masks) and if CAD-based layout modifications do not impact device models. Design DfM by CBC is the most deterministic approach but potentially has the highest impact on the product TTM.

For legacy technologies, the aggregate cost of MfD (machine, manpower, risk) is low due to the amortized and proven equipment. For new technology nodes, the increasing cost of MfD solutions may favor design DfM by CBC as the primary and CAD/design DfM as the secondary solutions.

4.2 Lithography

For planar technologies, for which the layers of different properties are subsequently deposited on top of the wafer, accurate reproduction of the layout pattern onto these layers remains the key challenge to manufacturability.

The workhorse patterning methodology (i.e., photolithography) is facing a widening disconnect between the wavelength of the patterning light and the feature CD it is required to print (Figure 4.1). A limited number of discrete wavelengths, such as 365, 248, 193, and 157 nm, are being relied on to support a continuous spectrum of printable dimensions, from 300 to 22 nm. EUV and nanoimprint, one with a wavelength of 13 nm and the other one with a equivalent wavelength (resolution) of about 1 nm, can be considered urgent attempts to break away from the hardship of the patterning wavelength being much larger than the minimum feature CD and to reset the optical DfM by replacing it with MfD. Notice that in that pursuit, the 157-nm development was dropped [4], as still subject to the wavelength handicap, even if of a somewhat smaller magnitude. But, searching for non-photo-based patterning techniques is only one of the options for ICM, which is able to generate profits and create new products even with the preexisting technologies, without further scaling, but by optimizing for cost reduction. The technical background of photolithography should help find the trade-off between DfM and MfD for future development.

It has been known since the 1990s that one can define subwavelength features with accuracy good enough for volume production. As mentioned in chapter 2 the direction of the waveform of the transverse electromagnetic (TE) wave is perpendicular to the features on the wafer surface, making the relation between the wavelength and the feature size on the wafer of secondary importance. Because the photoresist resolution is actually limited by the refraction of the light, one can achieve adequate contrast to support selective resolution of the photoresist required by the etch process. Computationally intensive (DfM) illumination options (source mask optimization [SMO] [7] and inverse lithography technology [ILT] [8]) are able to control the refraction.

While SMO and ILT may help ICM live with the subwavelength gap, alternative pattern transfer techniques have been long considered to reduce it. X-ray lithography, developed in the 1980s [9], rebranded by the year 2000 as EUV, is a major MfD investment [10]. Another contesting technique is the NIL [3], which does not use the light at all. Their properties and prospects can be compared, based on the redefined Rayleigh formula for cost of resolution.

4.2.1 MfD Cost of Pattern Resolution

To improve the resolution of optical imaging, values of almost all factors in the Rayleigh formula for line width resolution R can be modified for the subsequent generations of imaging tools:

$$R = k_1 \frac{\lambda}{NA} \frac{1}{n_1} \cdot \sin \theta \qquad (4.1)$$

Parameter	Explanation	Development Trend
R	Resolution or Rayleigh's constant	Continuous reduction required for shrink pott
λ	Wavelength of the incident light	Digital reduction: 365, 248, 193, (157), 13 nm
NA	Numerical aperture, $NA = n \cdot \sin \theta$	Continuous increase: 0.6 to 0.95 (max = 1)
θ	One-half of the incident angle of the light beam on the stepper lens	Up to 90°
n_1	Index of refraction	1.0 to 1.7
k_1	Resolution enhancement factor depending on the properties of distributed light source	1.0 to 0.3

While DfM efforts related to layout processing focused on the k_1 factor, MfD development involved hardware upgrades increase NA, θ, changing the light propagation environment to increase n_1, and of course, reduce λ. It also optimized the properties of the photoresist to improve the contrast at nominal CDs measured as postdevelop photoresist slope at 50% of its thickness as a process-based cause of CD variation in locations of suboptimal layout.

One can propose to substitute coefficients in Rayleigh equation by the ones that express the cost of lithographic resolution instead of the resolution itself, based on the cost coefficients for the DfM and MfD approaches. Subsequently, one can suggest approximate formulas correlating the cost trend with the deliverables of the DfM or MfD effort. Equation 4.1 for standard photolithography would then be rewritten as

$$C_R = C_\lambda \times C_{NA} \times C_{k1} \times C_{n1} \times C_{\sin\theta} \tag{4.2}$$

Symbol	Cost Component	Proposed Approximate Cost Trend
C_R	Resolution	All trends Combined
C_λ	Wavelength reduction (EUV, NIL, e-beam)	$1/\lambda$
C_{NA}	Increase of numerical aperture	NA^2
C_{k1}	k_1 reduction by rule-based (RB) and model-based (MB) optical proximity correction (OPC), ILT, SMO	$1/k_1^2$
C_{n1}	Increase index of refraction (immersion)	$\exp(n_1)$

By calibrating the cost equation to the process in question, one can choose the best RoI strategy for DfM and MfD. Assuming that the product line can justify increasing the cost of resolution from C_R to C_{R2}, to enable development product by technology upgrades, by a factor of α:

$$C_{R2} = \alpha \times C_{R1} \tag{4.3}$$

C_{k1} and C_λ are the only factors expected to increase because C_{NA} and C_{n1} are already close to their physically supported limits. C_{k1} can be adjusted gradually depending on the aggressiveness of SMO and ILT OPC solutions, as opposed to C_λ, which requires a rapid, multifold cost increase (about 5×), due to the lack of applicable wavelength in the range between 157 and 13.5 nm. One can also expect that C_{k1} would be reduced over time due to the simpler data processing.

As C_R became a limitation for the profit margin, alternative technologies not supported by the optical process came under consideration as attractive ones in the acceptable price range. For example, the nanoimprint should be considered as a process with $C_R = C_{NIL}$ (Cost of Nanoimprint Lithography) outside the scope of the other competing techniques. Because NIL is a wholesale alternative to the existing, single-mask/single-source lithography, its cost components cannot be directly compared to the ones for the other lithography techniques but requires their own equation similar to 4.2.

4.2.1.1 Extreme Ultraviolet Lithography

Extreme ultraviolet lithography (EUV or EUVL) is the best-known MfD solution for pattern transfer [11]. It uses extreme ultraviolet light with a wavelength $\lambda = 13.5$ nm and multilayer, absorber (purple) masks for image formation (Figure 4.3), such that the reflected EUV radiation is absorbed in

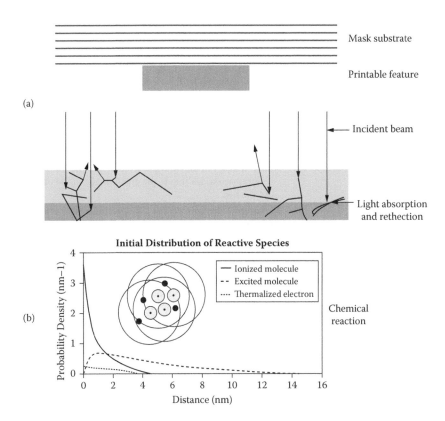

FIGURE 4.3
EUV image formation and absorption. (a) light paths in the photoresistor; (b) probability density of molecules or electrons impacting process resolution. (From Wikipedia.)

the resist and the substrate producing photoelectrons and secondary electrons. The random secondary electrons isotropically enhance chemical reactions in the resist, resulting in loss of resolution and line edge roughness.

While EUVL is a significant departure from previous generations of lithography with a long list of new technical issues (Table 4.2), it is still subject to the Rayleigh resolution equation, which seems to be the overpowering factor in its favor over the competing techniques or their physics. As all matter absorbs EUV radiation, EUVL needs to take place in a vacuum, which adds complexity, but presents no unknown risk to optical elements or their physics, including the photomask, which now must use defect-free Mo/Si multilayers to reflect light by interlayer interference, making the EUVL system absorb 90% of the light. High mirror absorption can be avoided in maskless interference lithography, but it is restricted to producing periodic patterns. While requiring new layout architectures can open up an opportunity for DfM in EUVL-specific designs, one can expect periodic patterns to be printable with less-advanced techniques as well.

TABLE 4.2

Key EUVL Technical Issues

EUVL phase defect	Defect reduces the intensity in the dark region by over 10% due to scattering by the 45-degree phase shift. The relative amount of scattering from phase edge defect increases with the amount of phase error.
Positive charging	Photoelectrons ejected from the top resist surface may cause damaging electrostatic discharge.
Deposition of contamination	Ambient or outgassed hydrocarbons deposit on PR.
PR etching by oxygen, argon	Ambient gases in the lithography chamber are ionized by EUV radiation, leading to plasma damage of the optics and the PR.
Printability	Minimum size of printable EUV defect increases with decreasing defect height. Defects 0.3 nm high are still significant. Multilayer defects over or under the layered structure are particularly insidious. The phase shift caused by a 3-nm mask flatness variation could produce a printable defect.
Reflectivity	Defects tilt the overlying multilayer mask, resulting in loss of reflectivity. A 1-nm deviation from flatness may lead to a 20% reduction of the image intensity [11]. Single, 200-nm wide atom high defects can be printable.
Shot noise	Poor control of exposure dose for features < 40 nm. The required sensitivity of 5 mJ/cm^2 implies only several thousand EUV photons accumulate in small areas. Increasing the dose will reduce the shot noise but increase the flare due to more free electrons.
Secondary electron blur	Secondary electrons due to random EUV dose variations reduce image contrast and resolution (blur), require X-ray absorber [25] (erbium) over the resist and reduce low-energy secondary electrons. Proximity of photoelectron trajectories reduces the exposure tolerance, causing line edge roughness.
Flare	Up to 16% flare effects difficult to separate from the secondary electron effects.
Photoresist heating	EUV absorption causes heating; ~100 mJ/cm^2 would result in ablation. The point spread of EUV chemically amplified resists is ~20 nm from the absorption point, indicating an approximate 40-nm PR resolution limit.

EUV light must be incident at an off-axis angle (6°), and as a result, the mask absorber shadow projected down to the wafer is significant for 32 nm and beyond [11]. The mirror collecting the light is exposed to a powerful EUV-generating laser or discharge pulse plasma and therefore vulnerable to damage from the high-energy ions and other debris. The multilayer mirror reflectivity causes loss of resolution due to plane diffraction. As a result, the resolution is not expected to be superior to 193-nm double-patterning immersion lithography (DPIL), making EUVL more of a tactical MfD response than a significant improvement in technology capability. The wafer throughput of the vacuum-based EUVL exposure tool is limited to a few wafers/hour [11].

Another EUV manufacturability challenge is optimizing PR (photo resist) thickness. Resist as thin as the pitch would ensure more uniform absorption and reduce forward scattering from the secondary electrons, but the resulting small devices would be subject to increased irradiation by photoelectrons, creating reliability concerns.

While the nominal line resolution expressed by the Rayleigh equation seems to make EUV a major MfD improvement over standard lithography, an abrupt reduction of λ by 15× does not mean worry-free pattern definition. Design and layout enhancement techniques still need to account for the many process marginalities, such as loss of resolution due to energy absorption in the PR. It is unclear if properties of devices patterned with EUV would be satisfactory for the equation for the cost of resolution to produce RoI values better than those for standard lithography. A comparison of MOSFETs built with DUV and EUV lithographies (Table 4.3) indicates that the RoI is greater than 1 only for a process node beyond 22 nm, but the poor accuracy of device models based on EUV capabilities would deteriorate it even further.

It appears that EUV fails the first viability test, that is, Would it work if built? While due to engineering efforts the EUV prototyping phase is concluded, the third level of the DfM test, that is, Can we build devices at a profit? with a throughput of 4 wafers per hour (WPH) using a 120-W source, would need a 25× performance boost to reach the 100=WPH entitlement, with a 3-kW source not available in the foreseeable future. Based on failure mode and effect analysis (FMEA), EUV technical problems may be comparable to those of its direct competitor, NIL, but their occurrence and detectability can be lower due to the long legacy of light-based pattern definition. One can calculate the uncertainty (or risk) measured as the ratio of the maximum to the minimum value of the FMEA product to be more than three times higher for NIL than for EUV (Table 4.4).

With tool cost considered a high-investment MfD solution, EUV is most likely to survive only as a mainstream technology. There are multiple pattern transfer alternatives for advanced ICM to be run in limited volumes, such as direct writing and DPT (double patterning techniques), which could also be used to manufacture complex products (Table 4.5).

4.2.1.2 Nanoimprint Lithography

As a technology with limited prior history of successful ICM implementations for volume production, NIL is viewed with skepticism due to the wide range of associated risks, known and unknown. NIL has been used to fabricate devices for electrical (MOSFET, O-TFT, single-electron memory) optical, photonic (e.g., subwavelength resonant grating filter), and biological applications (sub-10-nm nanofluidic channels in DNA

TABLE 4.3

Early Development of EUVL Printability MOSFET Parameters for DUV versus EUV Lithographies

Time	Company	EUV Device Parameters	Standard DUV Alternative/Benchmark
1996	Sandia National Laboratories, UC Berkeley Lucent Technologies	NMOSFETs. current $L = 130$ nm I_{Dsat} about 0.2 mA/μm, $L = 100$ nm, subthreshold swing $S = 90$ mV/dec, gmsat = 250 mS/mm.	NMOSFETs: 0.94 mA/μm I_{Dsat}, 60 mS/mm gm_{sat} = $S < 90$ mV/decade.
2008	IBM and AMD, College of Nanoscale Science and Engineering (CNSE)	90-nm trenches in metal CD uniformity 6.6%; overlay 16–17 nm correctable to 6–7 nm; power 1 W, wafer dose of 3.75 mJ/cm^2; defects 1/cm^2	200 W required for volume production. Optical lithography limit has a significant advantage.
2008	IMEC	~60-nm contacts doses: 12–18 mJ/cm^2	
2008	SEMATECH Samsung and IM Flash Technologies	22-nm half-pitch with chemically amplified PR; at 15 mJ/cm^2, line width roughness 5–6 nm,.	Spacer double patterning for 30-nm NAND Flash
2009	IMEC Intel	22-nm SRAM cells, contact, metal 1 (45-nm design rule).	Double patterning (double exposure) viable to 11-nm node

stretching experiment), and so on [12]. The known risks are related to poor overlay, and their mitigation is new process architecture. The unknown risks and the early life problems may only emerge commitment (corresponding to the 80% investment point at the time of per the 80/20 rule). Because NIL creates patterns by mechanical deformation of imprint resist and does not involve any elements of traditional optical imaging, such as steppers or photoresists, to achieve high resolution, most of the litho-friendly design (LfD) strategy would have to be reinvented. Yet, it is considered a simple production process with low cost and high throughput or a good "plan B," even if incompatible with the existing solutions.

TABLE 4.4

FMEA Range Comparison between NIL and EUV

	Severity	Occurrence	Detectability	Product	Range Ratio
NIL	3–5	3–8	3–7	27–280	10.5
EUV	5–8	5–6	4–6	100–288	3

TABLE 4.5

Competing Lithography Enhancement Techniques

	Method	Drawbacks
Rule-based OPC	Compensate process bias with line bias	Does not address corner rounding/shape distortion
Model-based OPC	Add sublithographic features, typically to the corners (line ends) of drawn geometries, to compensate for rounding and pullback; data verified by simulation	Complex setup and models requiring CoL
Inverse lithography/ SMO	Calculate mask features for required wafer pattern Pattern calculation of illumination source	Mask geometries not related visually to final wafer pattern
DPT	k_1 reduction by adding a patterning step	Double the cost of critical marks, alignment issues
NIL	Physical patterning of resin	No history in ICM alignment issues
EUV	Reduction of λ in Rayleigh equation	Extreme cost due to litho/mask/ resist/optic problems
Ebeam direct write	Reduction of λ in Rayleigh equation	Low throughput

NIL, first proposed by Professor Stephen Chou and his students [13], has been added to the ITRS Roadmap for 32- and 22-nm nodes. Toshiba was the original company to validate NIL for 22-nm lithography [14]. The imprint resist (thermoplastic or photo) is cured by heat or UV light with controlled adhesion to the prepatterned template. Then, the mold with topological patterns is pressed together with the substrate and heated above its glass transition temperature such that its pattern transfers into it. Reactive ion etching can then transfer the pattern in the resist to the underneath substrate. Photo NIL (P-NIL), uses a photo- (UV-) curable resist and a mold made of transparent material like fused silica, while electrochemical NIL (E-NIL) uses a stamp made of a superionic conductor such as silver sulfide. When all the metal is etched away, the complementary stamp pattern is transferred to the remaining metal. A high-throughput, full-wafer NIL template uses isotropic fluid pressure (Air Cushion Press, [15]). In comparison, low-throughput, die-level, step-and-repeat NIL has better layer align capability. As NIL is a 3D process, imprint molds can integrate the design data from more than one drawn layer to build multiple layers of vertical topography, replicate them with a single imprint step, and reduce fabrication costs by improving throughput. The overlay issues may also be mitigated this way. A range of materials is available to eliminate the sacrificial etch-resistant polymers [16], and a functional material may be imprinted directly to form a layer in a chip with no need for pattern transfer into underlying layers, also resulting in cost reduction, but at the potential cost of developing new ICM process architecture.

Other than overlay, the known risks of NIL are defects, template patterning, and wear. The current 3σ overlay NIL capability is 10 nm, and it is better for step-and-scan than for full-wafer imprint [17]. Defects from the template smaller than the postimprint process bias can be neglected, but bigger defects would require template cleaning or the use of intermediate polymer stamps [18]. Template patterning can be performed by electron-focused ion beam lithography or by high-resolution optical patterning (e.g., DPT) or another low-throughput, but precise, process. The template wears off due to the substantial pressure to imprint a layer. Damage to the porous low-k materials on the IC substrate can also occur. The residual layer following the imprint process should be thick enough to support alignment and throughput at low defects. However, its thickness has a negative impact on CD control during the etch, which is similar to the develop process in conventional lithography. Combining photolithography and nanoimprinting would mitigate the problem, at the cost of process complexity.

Because NIL relies on displacing the polymer, it would not suffer from wavelength-related pattern distortion correctible by OPC. Instead, it shows pronounced mechanical proximity (i.e., dependence on location and distance). An array of depressions is quickly filled, resulting in better quality at the edge

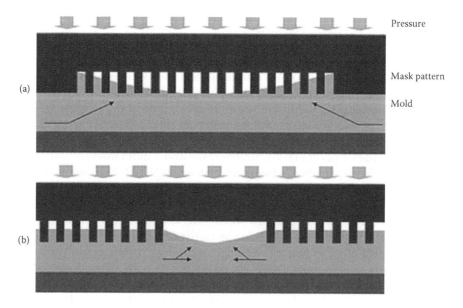

FIGURE 4.4

NIL cross section showing the principle of operation and pattern distribution. a) array of depressions more quickly filled at the edge than in center. b) wide space between protrusions. (After S. Laudis et al., *Nanotechnology.* 2006. 17, 2701–2709.)

than at the center of the pattern (Figure 4.4). Wide spaces between two groups of protrusions would be filled more slowly than the narrow spaces, forming holes in the unpatterned area. A large, dense array of protrusions will displace significantly more polymer than an isolated protrusion, and depending on the distance from the array, the feature may not imprint correctly.

To create DfM guidelines as part of the process setup for NIL, large-scale industry commitment is required. Nanoimprinting is built on a dramatically different process of physics and engineering compared to photolithography. It is not limited by diffraction or by scattering of secondary electrons, and does not require any sophisticated chemistry. As such, it would present a new set of DfM opportunities and challenges if used at a scope larger than the current laboratory scale. One of the enhancements it offers is 3D patterning of damascene interconnects and T-gates in fewer steps than required for conventional lithography. DfM efforts for NIL can be directed to

- proposing new design architectures to reduce process development cost and enable new functionalities
- developing pattern enhancement features to improve printing uniformity and reduce defects
- creating new tool sets for CAD support for device and process models in 3D

Template generation by self-assembled structures or double patterning may provide solutions for periodic patterns at 10 nm and less [19].

4.2.1.3 Double-Patterning Techniques

One advantage of DfM over MfD is that the DfM solutions are frozen into mask pattern and cannot be easily, or mistakenly, changed. In contrast, complex MfD process solutions may introduce extra risks, such as the

- risk of wafer handling—adding more manufacturing steps for MfD requires more handling of wafers and longer fab time
- risk of misprocessing, especially for complicated process recipes

But for simple experimental verification, tentatively adding manufacturing and masking steps may be a good trade-off to get quick access to information about whether it would it work if we build it. For some types of products, it may be impossible or impractical to meet all patterning challenges of some physical layers. This type of MfD represented by splitting one mask layer into three layers presents more risk than MfD requiring equipment upgrade, as each additional process step has a potential to introduce defects (i.e., lower the yield), though may present an overall higher RoI.

The key reason for mask splitting is to improve pattern resolution outside the original scope of the Rayleigh equation. To extend the capabilities

TABLE 4.6

DPT versus EUV Tool for the Future Process Nodes

		EUV Tool		Double-Exposure Tool	
Node (nm)	Year	NA	Depth of Focus (nm)	NA (@193 nm)	Depth of Focus (nm)
22	2011	0.25	107	0.93	408
16	2013	0.35		1.20	216
11	2015	0.35		1.35	148

of current 193-nm water immersion lithography toward the 22-nm node, double-patterning lithography is a promising option when contrasted to EUV (Table 4.6) It makes the k_1 factor no longer relevant, bypassing its theoretical limit of 0.25 by breaking down the different contributions to CD uniformity (CDU) and overlay. Beyond processes for memory products, lithographic patterning capability needs resolution better than that of ArF water-based immersion tools, even when approaching the practical limit of $NA = 1.35$. DPIL (double pattern immerssion lithography), an intermediate solution, splits the pattern at double pitch (i.e., 88 nm), which is used to get 22-nm features at 4-nm pitch [20]. The two images do not interact in the resist layer during the double-patterning integration flow. This raises new challenges related to mask design, mask manufacturability, and CDU. The two critical issues to be overcome are layout decomposition and overlay error.

Layout decomposition is a CAD process that requires placing selected features from previously a one-mask design layer such as poly to two new layers, such as poly lines and poly gaps. One should notice that in assisting with a mask split, CAD is not acting as DfM enhancement but as algorithmic support.

There are several DPT lithography processes; litho1-etch1-litho2-etch2 (LELE) [4], spacer-type DPT [4], and litho-oriented DPT [4]. They are all highly complex and involve multiple common challenges in both design and manufacturing sides, such as layout decomposition and stitch minimization.

Two basic process flows which can achieve pitch doubling through double patterning are: the dual-line approach and the dual-trench approach (Figure 4.5). In principle, both approaches can be used to pattern either a trench dark field (DF) layer or a line light field (LF) layer. The associated patterning steps have multiple implementation options, presenting a range of different requirements, capability, and limitations to meet resolution and control budgets. In a full manufacturing flow, DPT will require implementation of both process options to address pitch doubling on front-end and back-end layers.

The DPT options come with an incremental, overall process cost of up to 20% for the entire flow (Figure 4.6). For the dual-line process, a possible improvement consists of skipping the intermediate etch step by treating the first litho pattern such that the second resist layer can be coated and patterned on top of it. The design split of 2D patterns into two separate layers is an important

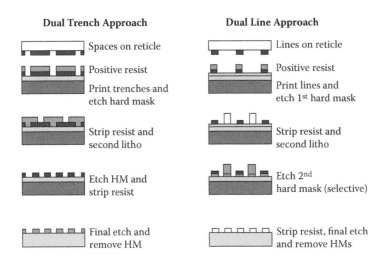

FIGURE 4.5
Wafer process flows for two DPT options.

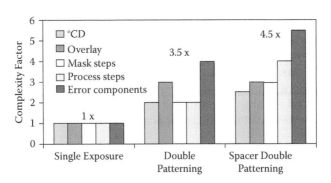

Type	Examples	Old process	New process	Cost factor	Old total cost	New total cost
Critical	Active, Gate, Metal1, Contact	5	9	1	5	9
Intermediate	S/D Implants, Metal X > 1, Vias	20	20	0.7	14	14
Simple	Well + VT adjust implants	5	5	0.4	2	2
	Total	**30**	**34**	**N/a**	**21**	**25**
	Additional cost	13%			19%	

FIGURE 4.6
Cost estimates of the different DPT options.

key for the success of double patterning, dependent on the pattern density, including 2D content, and on the application of the features to be patterned (conducting layers or contact holes, logical or digital circuits).

Intuitively, one would place alternating dense lines on mask A and mask B, thus doubling the pitch. A split like that to eliminate subresolution pitches and small spaces, while requiring the cutting of polygons, needs to ensure robust stitching through process variations. Design split rules for optimum cutting and stitching strategy have to account for scripting complexity and design regularity. For some applications, split conflicts arise that can only be solved by applying new design restrictions.

In a common double-patterning process (i.e., the double-line LELE, lines are printed and etched into a hard mask. This is followed by printing and etching a second series of lines onto a second hard mask or substrate. The final image is a combination of two sets of lines, and tight overlay performance becomes critical to avoid their misalignment.

Process challenges related to the existence of two CD populations: CD1 and CD2, require tighter metrology and CD control. To measure the overlay inside the double-patterning structure, CDU and overlay are both critical. Different etch bias for CD1 and CD2 demands precise respective litho1, litho2 targeting, and CD control at every litho and etch step.

To enable optimization of the total overlay control, modeling of mask registration error between the split layers as well as of tool compensation and process-induced systematic layer-to-layer fingerprint is required. After correcting for interfield (wafer) distributions, the simulated residual fingerprint indicates that an overlay of better than 4 nm could be achieved. Cost-effective process flows, automated mask design splitting, CD and overlay control, and metrology are the critical challenges to meet manufacturing requirements and make it a reproducible process with a CDU of 2.5 nm.

A detailed routing algorithm for DPT to improve layout robustness against overlay error would minimize indecomposable wire length and the number of stitches.

DPT is a generic application for poly, metal, active, and even via layers; current manufacturing infrastructures (e.g., stepper) and materials (e.g., photoresist) can be reused without expensive modification.

Unfortunately, decomposition is not always feasible, especially for complex 2D metal layers [24], owing to spacing constraints. For indecomposable cases, a conceptually simple but expensive solution is to modify the layout. Another solution is to split one polygon into two to resolve decomposition conflicts, which will introduce a stitch (Figure 4.7). A stitch is highly sensitive to overlay error, potentially causing pinching or bridging issues [25].

Most hard-to-decompose patterns are from complex 2D routing wires, but there is considerable design flexibility to find reasonable trade-off between DPT and conventional design objectives (e.g., timing, via, wire length).

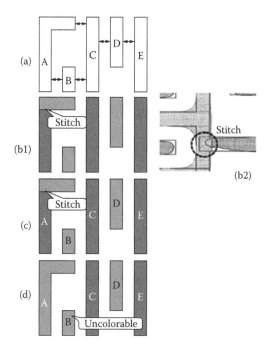

FIGURE 4.7

The concept of a stitch and layout decomposition based on a two-shade conflict graph (a). The layout cannot be decomposed by the two-shade approach (b1, c, d) for A, B, C, D, E- layout features independent of how to split nodes. Therefore, layout decomposition for DPT is not equivalent to but much more complex than the two-shade approach. b1, b2 show the potential stitching problem. (From M. Cho et al. 2008. IEEE ICCAD, p. 506. With permission.)

A polygon can be split to resolve a decomposition conflict at a cost of stitch. As mentioned, a stitch may result in significant printability degradation due to overlay error and line-end effect, to accomplish high layout decomposability and reduce the number of overlay-error-prune stitches. A DPT-friendly routing directly should output a decomposed layout without an extra time-consuming decomposition step.

Stitch minimization is therefore critical issue in DPT due to the overlay error, which is caused by the mismatch between the first patterning and the second patterning. A stitch is known to be highly sensitive to the overlay error, causing bridging, pinching, or notching (Figure 4.7b2). Layout decomposition and stitch minimization have to be considered during mask data preparation.

During layout decomposition, polygons will be first divided into two masks. Two polygons on the same mask should maintain a minimum double-patterning spacing. Such spacing sometimes enforces a specific placement for some polygon if there is another polygon nearby.

At the first glance, layout decomposition for DPT may seem similar to the phase assignment for a strong PSM mask [4]. However, there are two key differences. Phase assignment is for the space between polygons, but layout decomposition for DPT is for the polygons themselves. More importantly, resolving a conflict in phase assignment needs to involve layout modification (e.g., increasing spacing). Such that a polygon is severed into multiple polygons without altering a layout.

Consider splitting polygon A into two parts at a cost of stitch on A. Let us also consider the result of not selecting A in Figure 4.7. If we decide to split node B instead of A, it is still impossible to make the graph decomposable.

In general, the complexity of a layout decomposition for DPT with the minimum number of stitches is expected to be difficult, as there are many places for the stitches. Layout decomposition is the most critical step for DPT options, especially in metal layers, due to two-dimensionality of the patterns (while the poly layer has critical features in one dimension only). It can be very complex and cannot be solved by algorithms that require design time consideration, more specifically during detailed routing. A conceptual DfM-for-MfD flow would first require to finish detailed routing, then perform layout decomposition. If there is any undecomposable polygon, rip-up/rerouting should be performed repeatedly to fix the conflict, resulting in long design turnaround time [25]. A detailed routing oblivious to DPT may generate highly complex patterns, which may increase the wire length.

4.2.2 Overlay Road Mapping

Interlayer design rules, which pertain to enclosures or overlay of features located on different masks, should scale consistently with the CD rules of width and space to enable shrink path consistency. Equipment and mask makers understand that the overlay tolerance budget needs to be compatible with minimum CD reduction budget dictated by stepper resolution limits, mask minimum features for technology nodes, and overlay/enclosure (OL) between layers dictated by the mechanical tolerances of the printing hardware (Figure 4.8) [26]

One approach to evaluate if overlay tolerances corresponding to the mask and stepper parameters are acceptable for the technology node is by assuming the OL yield loss below an acceptable range (e.g., 0.2%) per mask level for 3σ misregistration error. While the analysis of equipment parameters shows improvement of overlay tolerance (OL) with CD reduction, any offset between the OL and the CD trends would need to be adjusted by matching of reticles and steppers in the fab and is becoming a critical MfD issue.

SRAM cell performance is a good metric of scalability of the different types of design rules (Figure 4.2). Correlation between CD rules of width and space and mechanical capabilities of lithography equipment expressed

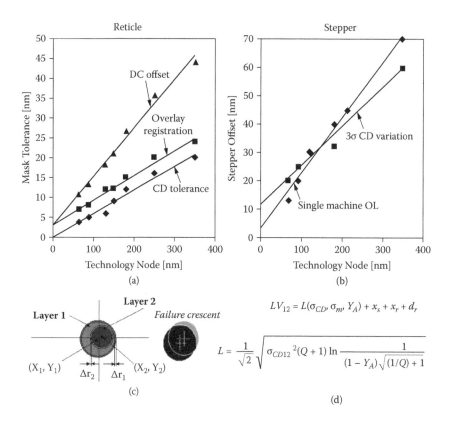

FIGURE 4.8
Mask and Stepper tolerance parameters vs. technology node defining design rule shrink path.
(a) 3σ CD and reticle overlay trends; (b) 3σ CD and "same" stepper overlay trends.

by the overlay rules was confirmed by two independent approaches [27,28], showing that the technology road map for equipment and mask manufacturing indeed put similar emphasis on the CD and OL shrinkability. However, the business aspect of providing tight OL solutions may in the future be more critical than the one related to the CD shrink path, as tighter OL may be required to take advantage of DPT in lieu of EUV or NIL.

To ensure low yield loss (e.g., below 10%) for a 40-mask process, CD and OL tolerances of steppers and masks need to reflect that the probability of geometries on two design layers overlap each other as required, is on the order of 99.8% assuming, normal distributions, without self-aligned processes. The correlation of alignment yield and maximum allowed misalignment can be described by the following formula [27]:

$$A = \sqrt{2}\left(B^2+1\right)^{1/2}\left\{\ln\left[\left(1-y_a\right)^{-1}\left(\frac{1}{B^2}+1\right)^{-1/2}\right]\right\}^{1/2} \qquad (4.4)$$

Parameter	Explanation
A	Cell area
B	Maximum allowed misalignment
y_a	Alignment yield

From this equation, B can be extracted knowing A and y_a and compared to the rule value required by the product profitability according to the technology road map.

An alternative approach to find the maximum misalignment value based on the assumed yield is to quantify cell cost as a function of overlay. Such analysis can be applied to the four-transistor cell, the area of which is expressed as a function of minimum feature size f_i and alignment tolerances a_i, impacting W/S/E (width/spacing/enclosure) rules for contacts and conducting layers. Assuming that

- f_i feature sizes for i critical layers in the cell are identical or closely related to the single value of minimum technology feature size f
- misregistration a_i on all layers is identical ($= a$), which can be accomplished by assigning all critical layers to the masks with the highest grade and running them on a single stepper

the area of a representative, four transister cell is given by [28]

$$A = 35\,f^2 + 88af + 48a^2 \tag{4.5}$$

For $a \ll f$, chip area and the die cost are given by (see Table 4.7 for explanation of equation parameters)

$$\text{Cost} = \frac{C_w \cdot A_0\left(1+r\frac{a}{f}\right)\left(1+\dfrac{D_0 A_0\left(1+r\frac{a}{f}\right)}{\gamma}\right)^{\gamma}}{A_w\left(1-\exp(-a^2/2\sigma^2)\right)^n} \tag{4.6}$$

The optimal overlay should be determined from the lowest cost point (Figure 4.9) as

$$\frac{d(\text{Cost})}{da} = 0$$

and calculated to correspond to a/σ in the range of 2.44–3.03. This means that the maximum allowed layer-to-layer misregistration in fact is close to the standard tolerance of 3σ, consistent for [27] and [28].

In summary, shrinking of CD and OL values with technology nodes is enabled by reduction of mask and stepper misregistration ($3\sigma_{CD}$ and overlay). Assuming the best single machine and mask overlay (the same stepper and

TABLE 4.7

Parameters Used in the Calculations of Overlay Tolerances

Name	Abbreviation	Parameter	Equation(s) or Figure
f	CD	Minimum feature size	Eq. 4.5, 4.6
a	m	Alignment tolerance	Eq. 4.6
σ_f	σCD	Standard deviation of minimum feature size	Fig. 4.5
σ_m	suffix	Standard deviation of misalignment or misregistration error	Fig. 4.8
$\sigma_{eff,l'}$ —		Variance of misregistration error	
A_c	A	Cell area	A_0*overhead*
A_0		Chip area with $a = 0$	
A_{chip}		Chip area	$A_0\left(1+r\dfrac{a}{f}\right)$
C_w		Cost of wafer	Eq. 4.6
D_0		Mean defect density	Eq. 4.6
γ		Learning curve parameter	Eq. 4.6
r		Area overhead factor	Eq. 4.6
n		Number of mask layers	Eq. 4.6
i		Layer number of a feature (e.g., fi, ai)	Eq. 4.5
y_a	ALY	Alignment yield	Eq. 4.4
R_l	R$_1$	Diameter of the larger feature in alignment	Figure 4.2
R_s	R$_2$	Diameter of the smaller feature in alignment	Figure 4.3
A	LN	Maximum misalignment, Lynch number	$A \equiv \dfrac{(R_1 - R_s)}{\sqrt{2}\,(\sigma_f/2)}$
B		Auxiliary parameter related to CD and misalignment variation	$B \equiv \dfrac{\sqrt{n}\,\sigma_m}{\sqrt{2}\,(\sigma_f/2)}$

mask to itself) for critical layers, such as active and poly, the tolerances scale linearly with technology node. The overlay offset trend, extrapolated to CD = 0, indicates of a few nanometers uncompensated CD tolerance and several nanometers of uncompensated machine tolerance which can be reduced by stepper tune-ups. Upgrading the fab process may challenge the overlay road map to scale down faster than the CD road map. DfM enhancements to design and process architecture should allow a wider use of self-alignment schemes, such as "poly first." Otherwise, new MfD approaches to enable advancements in pattern transfer (e.g., using dual patterning), or an abrupt change in the wavelength, which would allow for dramatic CD reduction, may force aggressive changes of alignment strategy. It is expected that future processes would

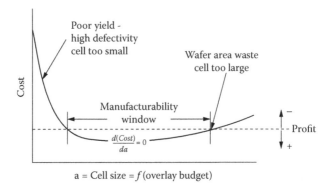

a = Cell size = f (overlay budget)

FIGURE 4.9
Impact of cell size driven by the overlay budget on die cost. Solid line is the cost equation; dashed line is the low but safe RoI depending on ASP (average selling price). High profit margins correspond to small cell sizes (risk too high), low profit margins to moderate cell sizes (medium risk), and losses to larger cell sizes (risk too low).

require more masking steps; therefore, layer-to-layer interactions have to be modeled with better accuracy.

4.3 Planarization

Chemical-mechanical planarization (CMP) (Figure 4.10) is a preferred process that can remove topography from physical layers in the IC: silicon oxide, metal, and polysilicon surfaces [29]. Scaling trends of MOSFET CDs required the use of CMP for an increasing number of applications, such as shallow-trench isolation (STI) and trenched metal interconnections (Cu damascene). CMP has also been utilized for fabrication and assembly of a microelectro-mechanical system (MEMS) [31].

In principle, CMP is a process of smoothing and planarizing surfaces with the combination of chemical and mechanical forces. It can be thought of as a hybrid of chemical etching and abrasive free polishing. Mechanical grinding alone may theoretically achieve planarization, but the surface damage is high as compared to when the chemical component is added. Chemistry alone, on the other hand, cannot attain planarization because most chemical reactions are isotropic. However, the material removal and planarization mechanism is much more complicated than just considering chemical and mechanical effects separately. CMP makes use of the fact that high points on the wafer would be subjected to higher pressures from the pad as compared to lower points, hence enhancing the removal rates there and achieving planarization [30].

CMP is most widely utilized in back-end ICM, where thin layers of metal and dielectric materials are used in the formation of the electrical

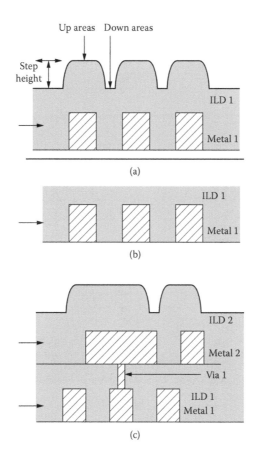

FIGURE 4.10
CMP process: cross-sectional diagram of internal dielectric on top of metal line (a) before and (b) after CMP; (c) multilayer structure. (From Zhengfeng, W., Ling, Y., Huan, N. S., and Luan, T. P. 2001. SIM Tech Technical Report Singapore Institute of Manufacturing and Technology. With permission.)

interconnections between the active components of a circuit (e.g., transistors, as formed in the front-end processing).

The interconnect is a thin film of deposited metal, selectively removed or modified in certain areas. A new level of thin film is deposited on top of the old film and the process is repeated until the interconnect system is complete. The goal of the CMP process is to planarize step heights caused by the deposition of thin films over existing nonplanar features, so that further levels may be added onto a flat surface (Figure 4.10) [4].

The damascene process, as well as its upgraded version dual damascene, is the critical technology in the transition from aluminum to copper interconnects in semiconductor manufacturing [4] to lower resistivity and increase electromigration resistance. In the copper interconnect fabrication, a simpler

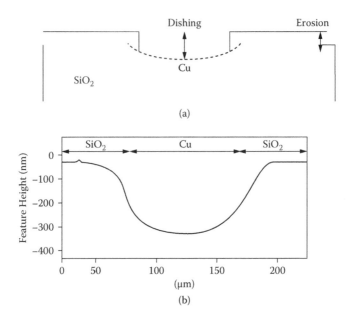

(a)

(b)

FIGURE 4.11
(a) Schematic representation of copper dishing and SiO_2 erosion. (b) Profilometer trace of a 100-μm line exhibiting 305 nm of dishing. (From Zhengfeng, W., Ling, Y., Huan, N. S., and Luan, T. P. 2001. SIM Tech Technical Report Singapore Institute of Manufacturing and Technology. With permission.)

dielectric etching replaces metal-etch as the critical step that defines the spacing of the interconnect lines, while the task of planarization moves to the metal deposition and CMP.

During the CMP of patterned copper wafers, copper dishing and SiO_2 erosion lead to deviations from the flat surface, during the overpolish step to ensure complete copper removal across the entire wafer (Figure 4.11).

Copper dishing is measured as the difference in height between the center of the copper line (i.e., the lowest point of the dish) and the point where the SiO_2 levels off (i.e., the highest point of the SiO_2). It occurs because the polishing pad bends slightly to remove copper from within the recess. The SiO_2 erosion is a thinning of the SiO_2 layer resulting from the nonzero polish rate of SiO_2 during the overpolish step, defined as the difference in the SiO_2 thickness before and after the polish step. Both copper dishing and SiO_2 erosion are undesirable because they reduce the final thickness of the copper line. Copper dishing leads to nonplanarity of the surface, resulting in nonuniform via resistance and parasitic distributions when adding multiple levels of metal.

CMP is also readily adaptable as an enabling technology in MEMSs fabrication, particularly polysilicon surface micromachining [32], easing their design and manufacturability by improving depth of focus and interlayer dielectrics (ILD) leakage due to severe topography. In general, CMP planarization alleviates processing problems of multilevel structures, eliminates design constraints of nonplanar topography, and enables integrating different process technologies. Design and DfM rules and models need to account for these differences.

A schematic cross-sectional view of a test wafer shows a copper layer of 2–3 μm to fill up the trenches of SiO_2. The adhesion layer under the electroplating metal is 100-nm thick. The dependence of the pattern density and line width versus polishing quality (i.e., the dishing in copper connections and the erosion of the surrounding SiO_2) is typically tested for lines and square arrays [33].

A one-head machine designed to polish semiconductor materials such as silicon, carbide, ceramics, metals, and brittle materials consists of a slurry supply system, pad surface condition, and temperature and condition detectors. A two-phase process adopted in inlaid copper pattern wafer polishing means changing slurry and working conditions during copper removal. The goal of the first phase of CMP is to remove the copper layer with high efficiency. This phase stops at the surface of the barrier layer. Then, the slurry and machine status are changed to the second phase for the remaining copper and barrier, with a removal rate at 1:1 selectivity. The benefit of a two-phase process is to eliminate the excessive dishing caused by oxidants used for a high copper removal rate. At the second phase, lower copper/barrier selectivity slurry minimizes dishing on patterned structures as well.

After CMP polishing on the patterned wafers under different head load and speed to characterize the polishing quality, the surface profiles are fetched through a profilometer and atomic force microscopy (AFM). A database of CMP process parameters, geometrical characteristics of the test wafer, and polishing qualities on the wafer surface (Figure 4.12) shows the relations between dishing (copper) and erosion (silicon oxide) and pattern characters (pattern density and line width) by the profilometer (>10-μm lines) and AFM (<10-μm lines) due to the limited scanning scope of AFM (100 μm in this case).

To derive CMP design rules by changing the process parameters (head load and spindle/carrier speed), relations between process conditions and surface finishing could be extracted to identify:

- The severest overpolishing of copper material (dishing effect; emerges, for example, when the head pressure is about 150 g/cm^2)
- The dishing depth of copper layer consistency with the machine speed to avoid speeds causing serious overpolishing of copper when the speed is, for example, around 75 rpm
- A possible trend between the erosion depth of the dielectric layer with the head pressure within probed process range
- Serious overpolishing of silicon oxide (e.g., when the machine speed is around 75 rpm)

From a macro view, overpolishing of copper (dishing) always occurred, but when lines are less than 10 μm wide, extruding copper lines above the dielectric layer dominate the situation. One explanation is galvanic corrosion

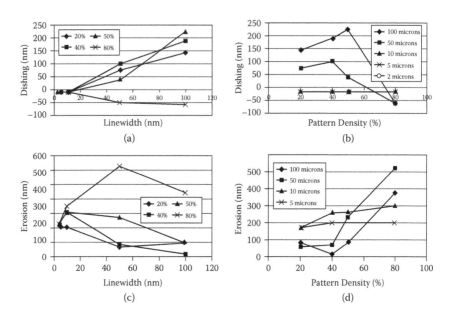

FIGURE 4.12
Correlations among: (a) linewidth and (Cu) dishing (with PD as parameter); (b) pattern density and dishing (with linewidth as parameter); (c) linewidth and erosion (with PD as parameter); and (d) pattern density and erosion (with linewidth as parameter), for a CMP process. $200 \, g/cm^2$ of head load and 50 rpm spindle/carrier is set for second phase polishing.

between the thin layer of titanium and the copper features, which attacks from the point when the titanium is first exposed and continues until all the titanium is removed. It can be said that dishing is occurring in the titanium layer rather than the copper layer.

Within-wafer uniformity of less than 10% of polished copper thickness determined by comparison of dishing heights on copper lines of different dice within the same wafer has been acquired.

CMP experiments with metallic and nonmetallic materials indicated that there are opportunities for process optimization that, depending on the type of IC design, can be more or less expensive. But, unlike for horizontal pattern transfer definition, there are no fundamental limitations for CMP to provide yielding silicon with sufficient DoE (design of experiments) on materials, machine, and setup. DfM is to expedite the efforts, but it is not the only viable option. For this reason, CMP design rules may differ between fabs running nominally the same process.

CMP is critical to replace the conventional etching process, which does not work on copper material. The two crucial defects of copper CMP technology, dishing of copper and overpolishing of dielectrics, can be reduced by collaboration between copper polishing and dielectric isolation, consumable development, and so on. With this, a variety of IC products, including opto-MEMS and bio-MEMS, will be the next possible areas of large commercial potential to explore, for which CMP can be the ideal tool for surface polishing.

To planarize the ILD on a wafer, double patterning with reverse masks may be a good low-cost option as an MfD supplement to DfM. For inductors or image sensors occupying a large fraction of the die area, it may be preferred to add a masking step to locally remove excessive ILD before the CMP. Unlike its high-precision DPT counterpart developed for line width reduction, DPT for planarization would require a low-cost photo process and a low-grade mask as it would only remove excessive ILD from a roughly defined region over the conducting layers.

4.4 Summary

In this chapter, we discussed trade-offs between MfD and DfM to achieve the best solution for high product RoI. In most cases, manufacturing can provide solutions to its own problems instead of delegating them upstream to modify design setup or execution. The only question is which approach is more efficient. As one example, splitting the manufacturing flow into more mask levels would improve

- device parameters by adding more implants
- printability by switching to DPT
- pattern density distribution by using reverse masks

But, these extra masks add manufacturing cost to every die, while the cost of CBC design and EDA solutions is divided by the product volume.

IC DfM is constantly facing many challenges and opportunities. It is not unusual for industrial engineering to approach a manufacturing "brick wall," which may prevent it from making further progress. Some forecasts claim that the end of IC manufacturing as we know it is always 7 years away, but eventually the industry has to undergo a transition to address ultimate limiting factors. If we look at the transportation industry, the saturation of the carrier speed was followed by both switching the carrier type and commoditization of the product and of its manufacturing technology. While the all-important questions of the day remain if technology A (that being, e.g., the EUV) would prevail over technology B (that being, e.g., the NIL), and whether that successful technology would be able to support itself in the long term. The overarching question should be if the market would be able to support more IC applications.

If the answer is affirmative, all the underlying questions are of secondary importance. The IC market, just like the transportation market before it, is gradually becoming commoditized: It has entered all aspects of life. But, the growth of demand for primary and processed information still seems to be far from over. While information is power in itself, it is also becoming easier to directly supply the power arm to the information and automatically run sensor-processor-actuator systems through the net of smart interfaces. To support this trend, ICM would need to become more flexible to cooperate with systems both at a micro- and a macroscale, not only with isolated solutions.

References

1. http://en.wikipedia.org/wiki/Multiple_patterning.
2. Kaiser, W., and Kuerz, P. EUVL—Extreme ultraviolet lithography laying the foundations for microchips of the next decade. *KGaA Optik & Photonik,* 2, 35–38, 2008.
3. Chou, S. Y., Krauss, P. R., and Renstrom, P. J., Nanoimprint lithography. *Journal of Vacuum Science and Technology B: Microelectronics and Nanometer Structures,* 14(6), 4129–4133, 1996.
4. Topol, A. W., La Tulipe, D. C., Jr., Shi, L., Frank, D. J, Bernstein, K., Steen, S. E., Kumar, A., Singco, G. U., Young, A. M., Guarini, K. W., and Leong, M. Three-dimensional integrated circuits. *IBM Journal of Research and Development - Advanced Silicon Technology,* 50(4/5), 2006.
5. Iniewski, K. (ed.). *Circuits at the Nanoscale: Communications, Imaging, and Sensing.* Boca Raton, FL: CRC Press, 2008.
6. LaPedus, M. Intel drops 157-nm tools from lithography roadmap. *EE Times,* 5, 2003.
7. Pomplun, J., Zschoedrich, L., Burger, S., Schmidt, F., Tyminski, J., Flagello, D., and Toshiharu, N. Reduced basis method for source mask optimization, BACUS Photomask Technology. *Proceedings of SPIE,* 7823, 78230E, 2010.
8. Zhang, J., Xiong, W., Tsai, M.-C., Wang, Y., and Yu, Z. Efficient mask design for inverse lithography technology based on 2D discrete cosine transformation (DCT). *Simulation of Semiconductor Processes and Devices,* 12, 49–52, 2007.
9. Bourdillon, A., Vladimirsky, Y., Morhoç, H., Nolamasu, O., Heaton, J., and Selzer, B. *X-ray Lithography towards 15 nm.* Jefferson Lab Technical Note 03-016, report of a meeting held January 24, 2003, Thomas Jefferson National Accelerator Facility (JLab), Newport News, VA, 2003.
10. LaPedus, M. EUV tool costs hit $120 million. *EE Times News and Analysis,* 11, 2010.
11. Bakshi, W. *EUV Lithography.* New York: SPIE Press, 2009.
12. Truskett, V. N., and Watts, M. P. C. Trends in imprint lithography for biological applications. *Trends in Biotechnology,* 24(7), 312–317, 2006.
13. Chou, S. Y., Krauss, P. R., and Renstrom, P. J. Nanoimprint lithography. *Journal of Vacuum Science & Technology B: Microelectronics and Nanometer Structures,* 14(6), 4129–4133, 1996.
14. LaPedus, M. Toshiba claims to "validate" nano-imprint litho. *EE Times News and Analysis,* 10, 2007.
15. Gao, H., Tan, H., Zhang, W., Morton, K., and Chou, S. Y. Air cushion press for excellent uniformity, high yield, and fast nanoimprint across a 100 mm field. *Nano Letters,* 6(11), 2438–2441, 2006.

16. Chu, K. S., Kim, S., Chung, H., Oh, J.-H., Seong, T. Y., An, B. H., Kim, Y. K., Park, Do, Y. R., and Kim, W. Fabrication of monolithic polymer nanofluidic channels using nanowires as sacrificial templates. *Nanotechnology*, 21(42), 425302–425307, 2010.
17. Bhushan, B. *Springer Handbook of Nanotechnology, Technology and Engineering*. New York: Springer, 2007.
18. Heidari, B., Löfstrand, A., Bolmsjö, E., Theander, E., and Beck, M. Pattern replication with intermediate stamp, U.S. patent 7704425, April 27, 2010.
19. Shevchenko, E. V., Talapin, D. V., Kotov, N. A., O'Brien, S., and Murray, C. B. Structural diversity in binary nanoparticle superlattices. *Nature*, 439(7072), 55–59, 2006.
20. Rokitski, R., Ishihara, T., Rao, R., Jiang, R., Haviland, M., Cacouris, T., and Brown, D. High reliability ArF light source for double patterning immersion lithography, optical microlithography XXIII. *Proceedings of SPIE*, 7640, 76401Q-1-7.
21. Wiaux, V., Storms, G., Cheng, S., and Maenhoudt, M. The potential of double patterning immersion lithography for the 32 nm half pitch node. Vol. 8, 2007. http://www.euroasiasemiconductor.com.
22. Chiou, T.-B., Chen, A. C., Socha, R., Kang, H.-Y., Hsu, S., Chen, H., and Chen, L. Full-chip pitch/pattern splitting for lithography and spacer double patterning technologies. *Proceedings SPIE*, 7140, *Lithography Asia*, 71401Z, 2008.
23. Cho, M., Ban, Y., and Pan, D. Z. Double patterning technology friendly detailed routing. *Proceedings of the IEEE/ACM International Conference on Computer-Aided Design*, 506–511, November 10–13, 2008.
24. Yuan, K., Yang, J.-S., and Pan, D. Z. Double patterning layout decomposition for simultaneous conflict and stitch minimization, *ISPD*, 107–114, March 2009.
25. Kahng, A. B., Hong, C.-H., Xu, P., and Yao, H. Revisiting the layout decomposition problem for double patterning lithography. *IEEE Transactions on Computer-Aided Design of Integrated Circuits and Systems*, 29(6), 939–952, 2010.
26. Allgair, J., et al. Taking aim at the overlay metrology budget for 70 nm. *Yield Management Solutions*, 29–34, 2003.
27. Lynch, W. T. The reduction of LSI chip costs by optimizing the alignment yields. *Proceedings of IEDM*, 7G–7J, December 1977.
28. Rung, R. D. Determining IC layout rules for cost minimization. *IEEE Journal of Solid-State Circuits*, SC-16(2), 35–43, 1981.
29. Zhengfeng, W., Ling, Y., Huan, N. S., and Luan, T. P. *Chemical Mechanical Planarization*. Singapore: SIMTech Technical Report, Singapore Institute of Manufacturing and Technology, Process Technology Division, 2001.
30. Stine, B. E., Boning, D. S., Chung, J. E., Camilletti, L., Kruppa, F., Equi, E. R., Loh, W., Prasad, S., Muthukrishnan, M., Towery, D., Berman, M., and Kapoor, A. The physical and electrical effects of metal-fill patterning practices for oxide chemical-mechanical polishing processes. *IEEE Transactions on Electron Devices*, 45(3), 665–679, 1998.
31. Hesketh, P. J., Huges, H. G., and Bailey, W. E. *Microstructures and Fabricated Systems IV*. Electrochemical Society, Technology and Engineering, Pennington, NJ. 1998.
32. *Micromechanical Systems: Advanced materials and Fabrication Methods*. Richard S. Muller, Ed. Washington, DC: National Academies Press, 1977.

5

DfM Metrics

5.1 Introduction

To quantify the design for manufacturability (DfM) success rate, one needs tangible metrics [1]. The cost and confidence level of parameters to choose from, including all aspects of reliability, yield, and cycle time, vary by orders of magnitude. The simplest ones are design rule checks (DRCs): A mandatory clean DRC report policy should ensure, at least theoretically, no fallouts until the end of the specified product lifetime. Manufacturing yield verification is hundreds of times more expensive and statistically less significant, as it is not possible to test every product for every aspect of its functionality. The situation is even more difficult at product qualification, for which small sample sizes and accelerated testing allow for very limited confidence in the validity of the data. For these reasons, physical DfM metrology is supported by conceptual parameters, such as criticality, occurrence, and detectability for failure mode and effect analysis (FMEA). FMEA enables problem anticipation and solving with only limited support of the actual physical data.

DfM metrics should be quantifiable within the design cycle and efficient enough to ensure high profits and easy product debugging. Quantification of design usually starts at layout, the first phase when product architecture is reduced to the expected silicon pattern. Design features can be grouped into a few categories (Table 5.1) and their key properties identified at product definition and confirmed at yield analysis. While reducing device variability by adding design constraints may increase die footprint and product development cycle time, causing loss of revenue, medium- and long-range pattern interactions (across-die or exposure field interactions) critical for product performance but, unlike short-range interactions not fully checkable by a DRC, require that design groups using different drawing techniques, whether manual or tool assisted, adhere to the same quality standards.

The first type of DfM metrics can be a summary of statistics of the final physical and electrical verification reports. Even with a zero-defect tolerance, layout techniques depend on the types of circuit elements, and blocks and design should extract the number of conflicting geometries, status of suboptimal DfM solutions, the number of device types, and nonadherences to DfM

TABLE 5.1

Matrix of Layout Interactions and Risk Mitigations for Product Engineering Issues

Domain	Type	Optical/ Pattern	Physical	Electrical	Time	CAD
Single feature	Issue	Can we print it?	Would it work if built			Efficiency of final verification
	Single layer	CD line, space, area of the photoresist	CD of printed pattern	Oxide reliability		
	Multiple layers	Enclosure, overlap	Contact fill	NBTI		
Feature-to-feature interaction	Issue	Can we manufacture it at profit?			Final verification schedule	
	Interaction	Horizontal	Vertical		Time dependent	
	Short range	RB OPC MB OPC	Stress relief	ESD latchup	OPC algorithm Runtime issues	
	Long range	CD control for etch	Planarity control for via resistance	Power domains	Fill pattern algorithm runtime issues	

"recommended rules" to correlate with product behavior at test and qualification. Layout verification log files should include a summary of outputs from the techniques used for the individual particular blocks and routing: manual, PNR, pCells (parametric cells), CBC (correct by construction), and so on. The verification summaries should be optimized from the standpoint of effort and TTM (time to market) to be the guidance for the next products. Problems encountered with device extraction and matching should also be captured. One can then discuss DfM metrics as a hierarchical approach, depending on the interaction distance in layout and length of the feedback loop in time.

5.2 Layout Quality

5.2.1 Variability

At the lowest hierarchical level of design, metric data are the parameters of layout primitives. Their impact on product performance and DfM complexity is the strongest for an active device (metal oxide semiconductor field effect transistors, MOSFETs). Their short-range features (critical dimensions [CDs], enclosures, and optical proximity correction [OPC]) are usually built into the standard DRC decks, but they do not guarantee matching of two nominally identical devices if verified at a higher (block or die) level. The implant layers, considered noncritical due to large CDs, may also give rise to device-to-device offsets caused by the differences in their neighboring topologies. Therefore, variability reduction metrics for active devices should be based on the checks and models of medium-range interactions (e.g., pattern density characterization depending on the window size and shape recognition). The ultimate goal for pass/fail criteria of sensitive circuits is matching on the extracted device properties with their models.

For the next category of basic layout components, passive devices (capacitors, diodes, resistors, inductors), designers may have to model with lower accuracy, with parameters assumed to be only a function of material properties and nominal layout CDs, without correcting for process variations or pattern reproduction inaccuracies. However, due to their larger size, passive devices may not only suffer from, but also be the source of, pattern reproduction and pattern density issues, depending on their medium- or long-range proximity.

The interconnects, the next level of layout features, are placed and extracted at the block level, but their aggressive CDs can add resistive or capacitive paths, requiring extensive modeling for radio-frequency (RF) applications. Other mask geometries created in the postdesign computer-aided is not (CAD) processing, such as fill pattern and OPC, are left out of the electrical design flow, and their electrical impact, although potentially nonnegligible, evaluated. The quality metrics of passives and interconnects would require die-level extraction and RF simulation (e.g., for the inductor Q factor).

5.2.2 Feedback Loop

Another quality metrics of layout DfM is the complexity and length of feedback loop from design to CAD to manufacturing (discussed in Chapter 3, Figure 3.18), which is prone to creating performance surprises. Feedback of yield problems to design requires first retrofitting of the design database, after all verifications, with delays and expense to the product cycle. Postprocessing at the mask level to enhance the layout for horizontal and vertical CD control (OPC and fill pattern), which may affect the parasitics but would not be retrofitted to design for extraction and verification, is the first source of poor determinism. This complex but nonqualified information is passed on to the mask shop, which does not have the information to understand the functions of the particular geometries but only relies on the CD control and defect inspection.

5.2.3 Critical Area

One layout parameter to define the susceptibility to the particle defects is the critical area (CA) [2]. The CA for shorts is the area for which, for a given radius of a particle, the center of a particle could land and cause a short in two wires (Figure 5.1).

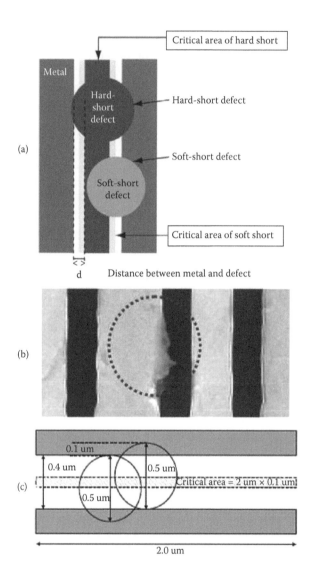

FIGURE 5.1
Critical area analysis (CAA): (a) Schematic of object types: Hard-short vs. reliability (soft-short failures (b) SEM image of suit-short failure after T. Ouchi, EDA Tech Forum, DfM Special Edition vol. 7 issue 2 2010. (c) calculating critical area force 0.5 μm particle (From R. Aitken. DFM metrics for standard cells, *7th International semposium on Quality Electronic Design*, 491–496, March 27–29, 2006. With permission.)

CA is a monotonically increasing function of the particle size. Bigger particles, while less likely to occur, cause more defects, with their size modeled as an inverse cube distribution. Combining the CA with particle distribution is called *weighted CA*. A single CA value can be chosen, such as the 50th percentile defect size (i.e., half the expected defect particles are larger, and half are smaller). The extreme cells at either end of the curve should be subject to additional analysis: those with high CA to improve yield, and those with low CA to assess layout effectiveness.

5.2.4 OPC Friendliness

To improve yield, OPC can be considered part of library design. Certain structures are inherently vulnerable to optical defects; therefore, allowance should be made in their design to ensure that subsequent OPC treatment will be able to treat them correctly. DfM metrics for OPC should help resolve layout conflicts among its key goals, such as

1. Improved printability: how complex the required OPC geometries are and how to build models for them
2. Reduced mask complexity: requiring low number of address nodes
3. Reduced CA: requiring as low layout density as possible
4. CMP (chemical-mechanical polishing) optimization: requiring uniform, and possibly high, layout density

Parametric tables of die layout properties can be set up for all these variables (e.g., to compare the impact of increasing contact overlap for printability and the increased CA for shorts). To allow for numerical optimization of rules, weighting can evaluate relative compliance for areas with the minimum rule and the recommended rule. While minor changes to the cell layout can increase the weighted total score, further improving the values may require an increase in cell area. Building a fractional compliance metric [3] requires assigning weight to the various rules and manual analysis. For full comparison, the metric must be created postlayout, rather than interactively at layout time.

5.2.5 Other DfM Metrics

The solution cost, the severity of failure modes, the TTM, and the yield ramp-up time should help resolve the trade-off between DfM and manufacturability for design (MfD) for variability reduction (discussed in Chapter 3, Table 3.6), depending on the product family. Two conceptually identical circuits, an SRAM (static random access memory) cell and a differential pair (Figure 5.2), may function in a digital circuit but fail in the analog one due to the different product and circuit requirements and the related design issues requiring different responses from technology or design, as different layout quality metrics are given for both digital and analog product families. For digital designs, with clock frequencies in the single gigahertz range and devices driven "rail to rail," the impact of parasitics on signal integrity due to the distribution of layout properties across the wafer should not exceed

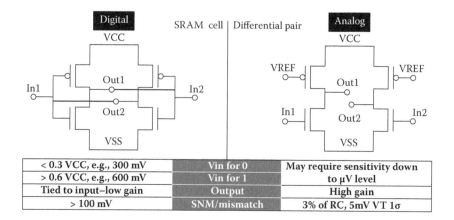

< 0.3 VCC, e.g., 300 mV	Vin for 0	May require sensitivity down
> 0.6 VCC, e.g., 600 mV	Vin for 1	to μV level
Tied to input–low gain	Output	High gain
> 100 mV	SNM/mismatch	3% of RC, 5mV VT 1σ

FIGURE 5.2

Complementary metal oxide semiconductor (CMOS) inverters forming the internal node of an SRAM cell or the input of an operational amplifier (VREF = reference voltage; In = input voltage; VT = threshold voltage).

the 100-mV noise margins or gain variations. These designs can use densely packed layout at the expense of accurate matching. But for similar output for RF designs operating in the frequency range of 100 GHz, the phase noise can make it impossible to realize analog blocks (e.g., the phase-locked loops). The sensitivity of operational amplifiers at the microvolt level requires parasitics simulation of all building blocks of the circuit. While digital design rules are still used in the verification, the unaccounted-for variation in parasitic capacitance and resistance can be above the sensitivity threshold driven by differences in CD transistor drive currents or contact/via resistances due to layout postprocessing. The consequential multiple-time variations of frequency response by multiple times can occur across the fab process lot, the wafer, or the die area, resulting in a degradation or loss of the primary function.

Among the DfM solutions for variability reduction, the benefit of rule-based approach over the CBC methodology is due to design options, enabling product development by adjusting device parameters and parasitics and creating opportunities to improve layout efficiency. But, in a rule-based approach, defining and verifying new rules is of increasing complexity due to the interactions among the integrated circuit (IC) geometries in spatial, electrical, and time domains (Table 5.1). One would identify critical devices, draw id layers around them, run a DRC to ensure that the devices exhibit the required behavior, transfer the data to the mask shop, write the mask, and proceed with device inspection based on the locations according to the id layers. Only when the id layer information is transferred to the fab to identify and characterize the devices on silicon can the full feedback loop from design to manufacturing be closed and included in the interaction model and solution cost. In comparison, the CBC approach is much simpler, not requiring information transfer.

5.3 Mask Ratio

One of the deliverables of IC DfM is to ensure that product launch will be done right the first time. But logical, electrical, and physical design for complex IC products (e.g., systems on chip, SoCs) may require iterations for model verification based on multiple mask sets. Fab yield may not be the best metric of DfM performance without considering reliability and interfab transfer complexity. For a foundry-based manufacturing, process development is largely design independent, and potential yield issues require adjusting the design without extensive DfM ↔ MfD interactions, as foundry processes are often not flexible and may have to low tolerance to suboptimal design.

To accurately calculate return on investment (RoI) of SoC DfM, one needs accurate device models, parasitics matching, uniform pattern density, and OPC to ensure high yield of the variety of SoC circuitry with different functions. Manufacturing yield as a metric may not capture all these issues due to the ambiguity of correlations between process variations and functional marginalities, depending on customer applications. A metric better suited than yield to indicate if design teams are adhering to DfM principles is the interfab product mask ratio (MR), that is, the ratio of all masks taped out in the product lifetime to the minimum set of the actual product masking levels.

$$MR = \frac{\sum Total\ Masks\ built\ during\ product\ lifetime}{Number\ of\ Fabs \cdot \sum_i Masking\ step(i)}, or \qquad (5.1)$$

$$MR = \frac{M_T}{M_0} = \frac{M_0 + s_{DfM} \cdot Y}{M_0} \qquad (5.2)$$

(See Eq. 5.3 for discusion of the parameters.)

Every mask change after the original tapeout would increase the MR beyond the entitlement value of 1.00. Meeting the entitlement of MR = 1 would mean that no extra masks (those not used in production) are ever built. MR > 1 would also mean a proportional increase of the development time, potentially having an impact on product marketability, as it reflects design cycle time increase and potential product performance issues.

Keeping MR close to entitlement needs to be first based on high quality of layout. In addition, the rule of 10 should help prioritize parametric correlations and their impacton yield, the cost of reticle plates, and impact of the design cycle time on product delivery related to debugging. With technology shrinks supplemented by other cost reduction measures such as foundry transfers, the MR becomes the key DfM metric to improve profit margins for the existing product and technology portfolios.

In principle, one should expect low MR to correspond to high initial yield. Integrating the design flow based on legacy DfM concepts (rule of 10, CBC, or variability reduction) should help drive down the MR, simultaneously improving yield [4]. Based on the rule of 10, one can propose three stages of DfM optimization for MR: definition, execution, and verification (D-E-V) (Table 2.2), each of which would have a 10-fold increased impact on the cost of layout quality. One can compare this schedule to the classical DfM P-D-C-A (plan-do-check-act) approach [5], which may suffer from a long cycle time if the first two phases consist of defining and running test silicon. Extending the D-E-V information feed-forward into the P-D-C-A feedback loop is only recommended when the expectation of the initial model accuracy is low and the risk spending time debugging an inherently nonmarketable product is high due to too many process and parametric unknowns. The P-D-C-A cycle softens the impact of the rule of 10 due to the existence of a feedback loop but adds nonrecurring cost (see test chip methodology, Figure 2.30).

At definition, the most critical but least-expensive stage of DfM implementation, the CBC methodology would help with design architecture at the single-cell level [6], for example, with layout tolerance bands or EPE set up to preserve design intent and ensure electrical performance [7] for modular design of multiport cells [8] or for die-level routing [9]. At the execution stage, 10 times less consequential but 10 times more expensive, design validation must identify tools for verifying or calibrating contextual intra- and inter-layer process proximity correction (PPC) [10]. Finally, at DfM verification, manufacturing feedback to design and layout would detect and correct local yield limiters (hot spots) [11], the step of local importance, but highest consequential cost if not done right, based on inspection or simulation using electronic design automation (EDA) algorithms or design kits [12]. Design data collection, storage, and transfer from IP (intellectual property) design houses to IC manufacturers (foundries) has to ensure efficient communication, cross-fab compatibility, and data protection, is also a part of DfM methodology, with efficiency measured by the cycle time.

5.3.1 Definition

To keep track and minimize MR throughout the three stages of implementation, a number of patentable original concepts have been developed. At definition stage, a technology library of CBC primitive cells may have tagged cell edges with self-created labels to enable matching layout parameters between the adjacent (abutting) geometries (Figure 5.3). The layout compiler would only place cells that meet DRC rules, by selecting a primitive cell from the library compatible with at least one previously placed instance, with edge codes providing the DR compatibility information. Geometries not complying with drawn design by themselves would resolve the conflicts when the cells are combined. The cells should have the same height and width, and once placed, the DfM algorithm would detect and correct the

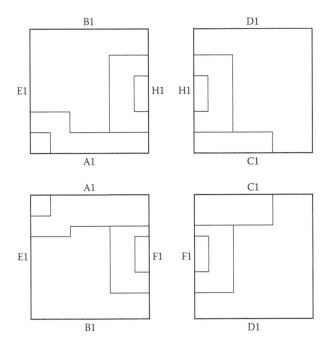

FIGURE 5.3
Integration of cells with edge codes (A1–F1) to ensure meeting design rules within a standard cell/technology library by CBC.

poorly manufacturable geometries undesired for lithography and etch, such as H-type, cranks, cross-links, microprojections, interdigitated combs, and o-rings [12]. On a manually synthesized layout, the statistics of the poorly manufacturable features can be the forecaster of yield.

5.3.2 Execution

One way to implement CBC in physical design is by modular layout of multiport cells integrated from single-port components [8], parsed (e.g., into read port, write port, and pull-up modules) for interconnecting and assembly. Unidirectional polysilicon and diffusion segments laid out perpendicularly to each other without breaks along the memory array inside multiport CBC bit cells would minimize process-induced nonuniformities between the individual MOSFETs. Interport interference would be further reduced by voltage lines shielding bit lines and word lines. By splitting gates, one can add ports to the cell at high array efficiency and increase the read performance of the port without increasing cell width. A read port may be split into "top" and "bottom" halves with a common word line connection, and the width or height of a bit cell would control the beta ratio. Poor quality leaf cells or their connections could have an impact on multiple masks, increasing the MR to two or three and extending product development cycle by a similar factor.

Another critical factor with an impact on MR is quality of design intent preservation throughout the manufacturing process. Here, one success measure is the conformance statistics of simulated silicon geometries to tolerance bands or the magnitude of edge placement errors (EPE). At the die architecture level, the coordinates of CBC cells and IP blocks have an impact on metal routing driven by the power supply requirements. The resulting die footprint can be optimized based on the cost metric [13] related to the putative chip performance and cost of manufacturing. For example, manufacturing cost can be a function of the length of routing lines and the number of their turns in a nonpreferred direction.

If design intent is not preserved on the wafer, correcting it may require multiple mask changes. Although the routing layers are usually restricted to top metals/vias, insufficient die area to accommodate DfM for routing can require an all-mask rebuild, increasing MR to two or more.

5.3.3 Verification: Model Accuracy and Hot Spots

The quality of IC manufacturing models depends on communication between design and fab and is another DfM parameter to be measured e.g. based on the number of process and product cycles of learning (CoLs). The number of CoLs can be reduced by converting process models into a set of kernels and then into a set of two-dimensional matrices [14] used as inputs to EDA tools. Creating a data matrix starts as a set of parameters (e.g., for lithography) stored in a database and converted into multidimensional kernels, depending on the individual aspects of the processing model. This is followed by the elimination of local yield limiters (hot spots), such as pinching, bridging, dishing, erosion, RC delays, metal line thickness variations, metal residues [15], due to either the poor circuit design or poor process control. Their detectability in the layout also prioritizes them for repair based on a score derived from interpolation with multiple risk factors and calculated from the slope for each of the hot spot types [14].

Lithographic hot spots can be classified based on three indices. The first index, sensitive to the exposure energy of the mask, can be obtained from the aerial image intensity curves. Its log slope (ILS) corresponds to a change in exposure energy relative to a change in CD. A normalized intensity log slope (NILS) is then derived by multiplying the ILS by the line width of the target line or space. The second index, sensitive to the image formation, is derived from the depth of focus of the width of the target line or space, based on a CD contour at a nominal, at a differential de-focus increments +df/−df, and its tolerance for the process window. The third index, sensitive to the mask manufacturing error, requires the knowledge of mask error enhancement factor (MEEF) based on the mask edge bias of the CD contour. One way to estimate this index is to add a small value (e.g., 2 nm) to each side of the layout line after OPC to capture the critical steps of the imaging process and their defect signatures.

Eliminating layout hot spots at DfM verification usually requires single-mask changes driving MR up to the 1.5 level. In agreement with the rule of 10, this phase has relatively lower impact on DfM metric setup, but the highest impact on the actual product performance.

Tight control of mask count and ratio also has its drawbacks. In planar technology, both horizontal and vertical CD control can significantly benefit, also to the advantage of yield, from adding mask layers. Also, device properties can be improved by adding implant masks.

Historically, three trends can be observed:

- Relaxing the criticality of masking steps by developing self-aligned process modules. Examples include Self-aligned source/drain implants that made it possible to relax alignment requirements of these masks to the gate. Self-aligned contacts allowed shrinking of memory cells.
- Adding more masks to drive new applications. Examples include improving field effect transistor (FET) (hot carrier injection, HCI) reliability due to lightly doped drain (LDD) implants, custom doping profiles for high-voltage applications, extra wells for soft-error latchup (SEL) reduction, and so on.
- Splitting mask levels to improve CD reduction or planarity control.

5.3.4 Correlation to Fab Yield

DfM implementation should result in high manufacturing yield, Y_{MFG}. At first glance, because the MR is impacted by the local design defects, its correlation to Y_{MFG}, impacted by process issues, may not be evident (Figure 5.4). However, it is a common practice to change product masks in response to DfM issues such as reliability or yield.

The total number of masks built to successfully run the product, M_T, is the sum of

$$M_T = M_0 + M_{DfM} \tag{5.3}$$

Parameter	Explanation
M_0	Number of masks in the minimum required product set
M_{DfM}	Number of extra masks changed to address DfM issues

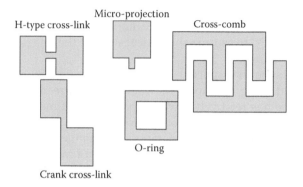

FIGURE 5.4
Poorly manufacturable layout geometries.

DfM yield, defined as the probability that the design would yield the first time, can be linked to MR as

$$Y_{DfM} = \frac{1}{MR} \qquad (5.4)$$

Therefore, fab yield improvement ΔY_{DfM} is related to the number of masks to be changed, M_{DfM}, as

$$M_{DfM} = s_{DfM} \cdot Y_{DfM} \qquad (5.5)$$

where s_{DfM} is the DfM yield sensitivity factor.

Assuming that the lowest detectable ΔY_{DfM} that can be attributed to a systematic design defect is 0.1% in order to trigger a mask replacement, that is, to cause MR > 1, the s_{DfM} factor needs to be 1,000. This factor, in turn, should be related to the wafer volume of the product:

$$s_{DfM} = k \cdot V \qquad (5.6)$$

Parameter	Explanation
k	Yield process sensitivity factor
V	Number of product wafers

Therefore, for the lowest detectable yield loss of 0.1%, a mask change may be required when product wafer volume exceeds 1,000. This, in turn, would have an impact on DfM yield:

$$Y_{DfM} = \frac{M_0}{M_0 + s_{DfM} \cdot Y_{DfM}} \qquad (5.7)$$

If the manufacturing process cannot be changed to improve yield, as is often the case for foundries, manufacturing yield loss would be equal to DfM yield loss:

$$Y = Y_{DfM} = Y_{MFG} \qquad (5.8)$$

ΔY may be due to the loss of reliability, functionality, or parametric performance. In accordance with the rule of 10, one may expect the sensitivity of reliability yield to be at 100×, of functional yield at 10×, and of parametric yield at 1×, compared to the nominal process sensitivity value s_{DfM}, because product marketability is by a factor of 100 impacted more by reliability issues than by performance marginalities (10 times) and parametric yield issues (one-to-one correlation). In other words, it is hard, but possible, to market a poorly yielding device, it is 10× harder to market one that does not perform well, but it is almost impossible (100× harder) to market a product that has a known reliability problem.

Accordingly, the MR would be more sensitive to reliability than to yield issues to ensure a much lower fallout level. While fab yield fallout of about 10% may be acceptable, functional yield loss for packaged product beyond 1% would be hard to justify, and reliability fallout of 0.1% of the population (i.e., 1,000 ppm) can be considered disastrous.

In the unlikely case of $\Delta Y \approx 100\%$ (i.e., the pre-DfM $Y \approx 0$), the MR may grow to a number $\gg 1$, corresponding to multiple updates of the mask set for products with high wafer volume or high yield sensitivity, identifying an unstable or poorly developed process or insufficient design rules.

One should also note that

- DfM yield impact is nominally lower when mask count increases because one mask corresponds to $\Delta MR = 2\%$ in a 50-mask set compared to $\Delta MR = 5\%$ in a 20-mask set. However, the cycle time sensitivity to MR increases with mask count at the same rate.
- DfM yield is inversely proportional to the initial suppression of manufacturing yield. This is because the higher the yield suppression is, the poorer the DfM will be.

IC performance and yield expectation related to DfM depends also on wafer volume. For low-volume parts, the MR may be kept at 1.00 as there is not enough yield data to justify design changes even with suspicions of failures, but one should avoid design flows where DfM can disturb performance of the chip, causing design respins.

While small yield improvement due to mask pattern enhancements (i.e., low DfM yield sensitivity) indicates a stable process, far off from a "yield cliff," it can also indicate that design rules are not aggressive enough to take advantage of the fab process capability. In other words, both too high and too low DfM impact on yield are questionable.

In conclusion, for DfM to increase the probability of design working the first time at high yield, the expected MR should be low. DfM yield as the inverse of MR is a function of the yields at all design stages: reliability, functional, and parametric, with the order of importance set by the rule of 10. Because DfM yield and MR are the metrics familiar to both manufacturing and design and available to them, they should be used in standard manufacturing yield models. Following DfM guidelines e.g., during circuit routing should help increase manufacturing yield by ~10%, which comes close to rebuilding 3 metal masks in a 30-mask process.

5.4 FMEA and RoI

Because DfM rules are not directly responsible for the functionality of the individual devices, but they address contextual layout issues (hot spots), the distinction between DfM and DRC rule decks has been the subject of many definitions. As the occurrence of hot spots increases with the number of product layouts, all design rules for mature technologies can be combined into one enforceable deck, but with cutoff lines decided by business factors. With the support of the FMEA and return on investment (RoI) methodologies, the criticality, occurrence, and detectability of failure modes and the associated cost of noncompliance would help define the enforceable rules [16,17]. Depending on the acceptable design risk level for SoC, logic, memory, analog, and mixed-signal circuitry, one can exclude the noncritical rules to save the implementation cost and time.

A distinction between DfM and DRC is a function of technology, EDA tool setup, and product application. Custom dummy pattern or OPC can be defined either separately and arbitrarily for the different regions of the die, potentially raising issues due to conflicting device properties at area boundaries, or by a universal flow, which first identifies layout properties in the different areas (e.g., dense, semidense, isolated) and then applies proper DfM routines to create the mask pattern. Metrics to judge the effectiveness of both approaches is device performance improvement by wafer pattern simulation, first optical, followed by electrical [18], as required by the product yield binning.

Phasing in DfM rules can be done at two levels. At the IP block level, standard cells complying to the basic design (DRC) rules, such as minimum feature dimension, spacing, and enclosure, would also require

- If placement-dependent scatter bars for deep-subwavelength printability
- slope and side lobe reduction for enhanced photoresist contrast
- shot count reduction for mask manufacturability: removing jogs, intersections, and other noncritical address nodes

DfM at the die level should include rules and algorithms to standardize the layout by

- eliminating forbidden pitches and ensuring proper line orientation
- equalizing local and global pattern density
- reducing microloading by proper spacing from small to big features
- managing the hierarchy for cell placement to adhere to the rules above.

Implementing DfM guidelines is associated with cost, and a metric is required to categorize their order.

FMEA [17] categorizes failure modes based on three criteria: severity S (or criticality), occurrence O, and detectability D, in 10 levels. The severity (Table 5.2) ranks failure modes as levels 10 or 9 when they "kill others" (i.e., create a situation dangerous to other products in line). Defects making parts nonfunctional are at severity level 8 ("dead on arrival"); the ones responsible for reduced functionality correspond to severity ratings from 7 to 1, with functional yield loss from 7 to 5 and parametric yield loss from 4 to 1. The occurrence ranks defects in percentage of the population or parts per million, depending on the acceptance limits. The detectability defines design responses for detectable failures and rates them depending on the detection complexity on the scale of 1 to 10 (Table 5.3). The FMEA metric is the risk priority number (RPN), that is, the product of severity, occurrence, and detectability (Figure 5.5).

$$RPN = S \times O \times D \qquad (5.9)$$

TABLE 5.2

Example of Severity Rating in DfM Rules

Effect	Criteria	Example	Rank
Hazardous without warning	Affects existing good parts or equipment	Photoresist lifting and redeposition on other wafers	10
Hazardous with warning	Affects existing good parts or equipment	Not possible to etch—no endpoint reachable	9
Very high	Current part inoperable	No useful PR pattern	8
High	Operable; reduced performance	Printing all lines not possible	7
Moderate	Operable at center of process window, low c_{pk}	Poor image contrast; potential for bridging	5
Very low	Operable, features visually inconsistent	Poor OPC correction of some critical lines	4
Minor	Operable; noncritical features off target	CD control poor along some connecting lines	3
Very minor	Minor/cosmetic items	Symmetry of fill pattern	2
None	No effect	Meets all process targets	1

TABLE 5.3

Example of Detectability Rating in DfM Rules

Detectability	Detection by	Example	Rank
Absolute uncertainty	No design control	Poor yield = High cost	10
Very remote	R, P, Q, T	Product failures	9
Remote	P, Q, T	Detectable at end of line	8
Very low	P, Q, T	Detectable at dedicated electrical test	7
Low	R, Q	Detectable by design review, custom ORC (optical rule check)	6
Moderate	Q	Detectable by custom ORC	5
Moderately high	Visual at defect site	Would be detected by standard ORC	4
High	Visual near defect site	Would be detected by DRC	3
Very high	Visual at block level	Gross pattern density issue	2
Almost certain	Visual at die level	Wrong data?	1

R = reviews; P = prototypes; Q = CAD QA, T = tests.

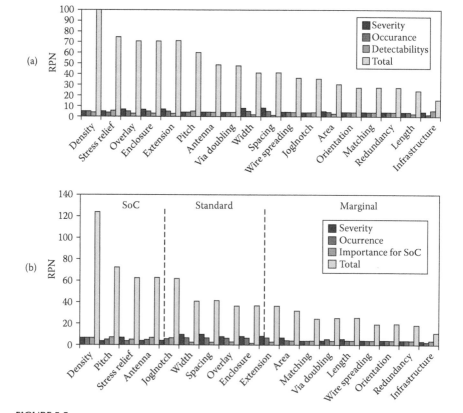

FIGURE 5.5

Risk priority number *RPN* as FMEA metric. (a) generic; (b) distributed between SoC and standard rules indicating also marginal failure nodes. (From A. Balasinski. 2006. *ISQED* pp 491–496. With permission.)

Depending on the quality expectation, one can truncate the rule implementation pareto starting from the failure modes with the highest *RPN*s. As the maximum $RPN = 10 \times 10 \times 10 = 1,000$, one can set up DfM implementation limit as a percentage of this value (e.g., 10%; i.e., $RPN = 100$) by enforcing design rules above such a cutoff line (Figure 5.5). The analysis needs to be done first for the IP blocks followed at the die level. This is because *RPN* numbers calculated separately for the individual types of circuits, such as memory, logic, or analog, SOC, would have their internal DfM quality metrics confounded by the matching issues of functional blocks with layout features of different pattern proximity, orientation, and density characteristics.

5.4.1 DRC/DfM Prioritization

Verification of DfM rules increases the cost of design. CAD tools are expensive and require time to learn, debug, and qualify for complex layouts. DfM may also increase the cost of mask writing and inspection by introducing features with nonstandard CDs, such as scatter bars. At the same time, cost of yield is increasing with every technology generation [19]. Therefore, RoI analysis for DfM should find the best trade-off between the cost and time of rule implementation and the cost of poor yield.

In general, RoI is be defined as a (unitless) ratio of money returned to money invested, according to the formula [20]

$$RoI = \frac{\$\ returned}{\$\ invested} = \frac{Net\ Income}{Book\ Value\ of\ Assets} \tag{5.10}$$

But justifing DfM through RoI (i.e., by comparing product value with and without DfM) may lead to substantial errors due to the many unknowns in the changes of *Net_Income* and *Book_Value_of_Assets* over time. An alternative approach to RoI is to calculate how much time it would take for DfM investment to be returned by the higher fab yield (i.e., simplifying the RoI tracking to a single point of time). This assumes that design adjustment by DfM is not affecting product parameters or properties; therefore, the RoI depends linearly on fab volume, which in turn increases with the time when the part is in production. The question is how fast would the time spent on DfM development be made up, and what the initial impact of schedule push-out due to the introduction of DfM rules is. That schedule push-out can be expressed as the cost of time, using a new parameter in DfM quality metrics, the time to RoI (RoIT).

5.4.2 Time to RoI

Time to RoI (RoIT) is defined as

$$RoIT(weeks) = \frac{Total\ CAD\ Cost(\$)}{(\ Yield - \ Area)(\$/wk)} \tag{5.11}$$

ROIT

Parameter	Explanation
Total_CAD_Cost	Combined cost of implementing DfM rules and of product revenue loss caused by the delay
ΔYield	Financial impact of yield loss incurred if the rule is not implemented
ΔArea	Financial impact of loss of silicon area due to the rule implementation (assuming that the die size has to increase due to device matching, orientation, or wire spreading)

The expected *ΔYield* can be modeled based on the severity, occurrence, and detectability from the FMEA analysis. The relationship between yield and occurrence is straightforward: The more often a defect occurs, the larger the yield loss. The correlation between yield and severity can assume that if the part is not yielding, it would not matter if the defect is severe or is just a persistent marginality disabling its primary function. To contain this ambiguity, one can combine the impact of defect severity and occurrence, such that yield loss is proportional to their product:

$$Yield = k_y \cdot [(S \times O)_{\text{without\_DfM}} - (S \times O)_{\text{with\_DfM}}] \tag{5.12}$$

where k_y can be called the defect financial sensitivity factor. Because yield is the detection mechanism itself, it may replace the detectability D in the RPN:

$$Yield = k_d \cdot D \tag{5.13}$$

where k_d is the defect detectability factor. Notice also that the $S \times O$ product before and after successful implementation of DfM rules should differ significantly enough for Equation 5.12 to be simplified to

$$Yield = k_y \cdot (S \times O)_{\text{without\_DfM}} \tag{5.14}$$

Assuming a Poisson yield model [21]:

$$Y = \exp(-A \cdot D_0) \tag{5.15}$$

Parameter	Explanation
A	Die area
D_0	Defect density

we can propose a formula to correlate FMEA parameters to defect density:

$$\ln(k_y(S \times O)) = -AD_0 \tag{5.16}$$

Next, we discuss two examples of RoI analysis for DfM rules of pattern density and layout simplification. To justify DfM rule introduction, the problem and the success criteria are first formulated, followed by CAD implementation and execution of the algorithm by design engineering.

Example 5.1: Making Pattern Density Uniform

Uniform metal pattern density across the die, accomplished by adding dummy fill geometries to design layout, should reduce via resistance by ensuring uniform thickness of interlayer dielectric (ILD). The solution path is to define the tile size and spacing from devices, connecting lines, and other keep-out regions and establish the execution sequence in design flow. As the fill algorithm has a number of desired features, such as short runtime, multilevel hierarchical structure, controlled impact on device performance, and its implementation could be complex enough to cause product schedule delays [21]. The impact of pattern density optimization can be first verified based on the $S \times O$ of the expected defects having an impact on product performance. Optimally, the fill process should be verified by chip-level simulation to include the added parasitic elements, mechanical, optical, or electrical.

The RoI model parameters include the failure severity factor and coefficients for yield, die area, and labor (Table 5.4). We assume that the original severity of high via resistance $S = 5$, which can be linked to the yield loss due to via failures. Because the failure mode would be present only in limited regions of the die, the occurrence rate without high DfM can be assumed to be $O = 5$, resulting in the $S \times O$ product of 25. Such $S \times O$ value should qualify the rule for priority implementation (Figure 5.5) with area penalty for the dummy fill to improve pattern density distribution over the large passive elements (capacitors, resistors) of less than 1%. Based on these parameters fed into Equation 5.11 (Table 5.5), RoI for the dummy fill was calculated for the *Total_CAD_Cost* expected from the required workload, and the final *RoIT* was found to be 9 weeks.

TABLE 5.4

Examples of RoI Model Parameters

Name	Unit	Value
Unit cost of labor	$/man-week	4,000
Wafer volume	wfr/week	100
Wafer value	$/wfr	5,000
Cost of delay	$/week	500,000

TABLE 5.5

Example of Pattern Density Rule Implementation: Impact on RoIT

Pattern Density	Effort (man-weeks)	Labor ($)	Delay (weeks)	Cost of Delay ($)	DPW Loss (%)	Total ($)
Design	4	16,000	2	1,000,000	0	1,016,000
CAD	1	4,000	1	500,000	0	504,000
Total CAD						1,520,000
Area penalty					5	25,000
Yield impact					−40	−200,000
Profit impact						175,000
Weeks RoI						9

One should note that the DfM metric definition may differ between the supplier (EDA) and customer manufacturing organization (FAB) due to their different hierarchy of interest (Table 5.6; [22]).

Example 5.2: Layout Complexity Reduction

Reduction of address nodes (or fracture polygon/shot count) in a layout database is a DfM measure that helps reduce mask cost due to simpler fracturing, shorter write time, and more efficient inspection. The opportunity to reduce the non-critical vertices arises first at the standard cell level. But, optimization of even a very simple layout feature, such as a metal landing pad connecting several contacts, has to consider different options (Figure 5.6). For example, a cross and a rectangular patch, while identical from the LVS standpoint, are dramatically different for DfM. The number of address nodes and the resulting count of edge placement errors (EPEs) is lower by a factor of three for the patch than for the cross due to the different corner count. The two shapes would therefore differ in design verification and mask writing times, and for a million cells in different layout environments, using the cross instead of the patch could slow down the tapeout by over 2 weeks (Table 5.7). The *RoIT* would depend on the comparison between the *Total_CAD_Cost* and the financial impact of tapeout delay.

TABLE 5.6

Fill Pattern Implementation Characteristics Showing Different Hierarchy of Interest to EDA and FAB

Responsible Group	EDA Preferences	FAB Requirements
Primary features	Runtime	Range (small)
Secondary features	Number of extra geometries (low)	Density (low)
	Extra capacitance (low)	Gradient (low)
	Reproducibility (high)	Overlay to other layers

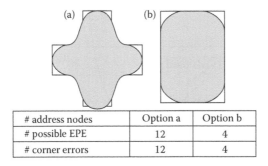

# address nodes	Option a	Option b
# possible EPE	12	4
# corner errors	12	4

FIGURE 5.6

Examples of parameterized cell complying with the same set of design rules but with significantly different DfM properties. (From A. Balasinski. 2006. *ISQED*. pp. 491–496. With permission)

TABLE 5.7

Impact of Layout Geometry on CAD Processing Time

Process Item Per Address Node	Total Runtime Per Address Node (Vertex or Shot) (Seconds)
CAD tool qualification and final runs	0.1
MB or RB OPC script cleanup	1.0
Mask write and qualification	0.5
Total schedule slip	1.6
Total per 1 million nodes	2.65 weeks

5.4.3 FMEA-RoI Correlation

The RoI of design rules depends on the failure modes the rules would prevent. Adding a design rule to ensure manufacturability may improve yield, yet the time to money may be longer than the time allocated for the new product development. Three scenarios may affect the time-dependent cost of rule implementation for high-, medium-, and low-severity failure modes (Figure 5.7):

- Scenario 1: (high severity) The new rule is required to ramp volume production (Figure 5.7a). (Standard, mandatory design rule)
- Scenario 2: (medium severity) The new rule helps develop the new product faster (Figure 5.7b).
- Scenario 3: (low severity) The new rule delays product introduction but improves its manufacturability (Figure 5.7c).

The incremental TTM due to design rule development makes the CBC approach a more attractive solution to reduce layout variability. Standardized layout based on parametric cells (pCells) with circuit elements predefined and characterized on test vehicles [23] would include all the required process-related geometries, such as dummy fill and OPCs. To assemble the die, they would be

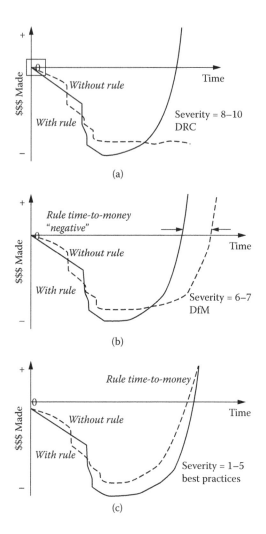

FIGURE 5.7
Three RoI scenarios with different RoIT and the associate severity rates: (a) rule is required, (b) rule improves TTM; and (c) rule degrades TTM but improves long-term ROI. (From A. Balasinski. 2006. *ISQED*. pp. 491–496. With permission.)

preferably laid out on a grid in a fixed orientation with the quality ensured at cell, block, die, and mask levels, also by checking design rules related to their environment. Modeling yield and turnaround for a CBC parameterized layout in comparison to the more iterative DRC development should involve, for example, the impact of minimum CDs and overlay tolerances, avoided for analog/RF design by either a rule check at the die level or building them into the layout structure. To parameterize the basic structure of a MOSFET (Figure 5.8), the design discipline has to be codified and observed (Table 5.8). The conservative analog/RF rules should recognize parasitic components of on- and off-die

TABLE 5.8

DfM Recommendations for Mask Manufacturability

Rule	DfM Recommendation	Mask Manufacturability Improvement
A	Avoid minimum printable CD values; acceptable CD variation depends on circuit application	Large CDs are easier to write, should rely on less-restrictive RET for better process margin
B	Make sufficient to avoid local channel length increase or pullback; keep fixed L across W	Reduces mask bias complexity for small features
C	Ensure no spacer encroachment but full placement on diffusion	Reduces impact of mask CD variation
D	Avoid proximity effect	Same as B
E	Ensure no corner rounding from implant layers, shadowing	Less aggressive mask OPC
F	Ensure no poly counterdoping	Same as C
G	Fixed number of contacts per unit L	Symmetrical contacts on both sides of FETs; alignment helps reduce mask address nodes, fracturing, and data file sizes
H	Avoid narrow width	Same as A
I	Ensure stable environment	Reduces at etch macro- and microloading
A/H ratio	Maximize area-to-perimeter ratio	Same as E

wiring, including vias and contacts; for example, to optimize gate resistance using the simulated f_T frequency as the quality metrics, one can identify the advantage of the double-contacted MOSFET gate (Figure 3.36), even if the CBC layout may increase die area by as much as 20%. The preference between adding design rules versus using a parameterized or fixed layout can also be decided based on the FMEA/RoI analysis.

Quality metrics for layout should include factors related to functionality, yield, and reliability, with parameters reflecting the sensitivity shown by analog/RF devices, requiring die-level verification of pattern density correction features, standby current control, and combined die and frame checks. To enable reticle-level data analysis to enhance wafer processing, dedicated mask DRC may also be required. For layout becoming the ultimate frontier of die quality, final verifications should focus on design intent, not just on pattern transfer. The regularity of the layout, such as the gate array matrix, improves circuit topology and parasitic.

Just like the FMEA, the RoI analysis would make it possible to prioritize DfM rules at block and die levels. Pareto distribution of rules (Figure 5.5) can be truncated depending on the product volume and the criticality for the technology family. Note that the longer the *RoIT*, the less critical is the rule implementation. The DfM, FMEA, and RoI rule rankings should correlate such that high *RoIT* (weeks) corresponds to a low *RPN* risk priority. Rule implementation requiring significant time investment but with low *RoIT* impact (e.g., for low-volume

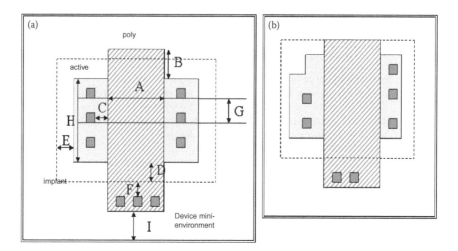

FIGURE 5.8
Sample layout of MOSFET realized with good and poor layout best practices.

products) include jog/notch reduction, data infrastructure management, and device matching. They are not critical at the initial stage of process and design development, such as for the test chips, but once the product line is in volume production, every wrinkle in the product flow may translate a meaningful revenue loss. DfM rule data points below the RoI trend line correspond to the moderate schedule risk, but their low development cost favors their implementation.

5.5 Summary

In conclusion, DfM metrics can be considered a hierarchical structure. At the bottom of this structure, there are layout statistics of EPE errors, nonconformality to tolerance bands, the number of undesired geometries, the count of mask data address nodes, and so on. At the second level, these statistics may be used as predictors for process problems manifesting themselves as hot spots due to the coincidental excursions of process parameters and quality metrics. The number of hot spots would now project on the third level of DfM metrology hierarchy, which is yield and mask ratio. Their correlation should help choose the solution path between DfM and MfD enhancements to resolve the layout problem otherwise driven by cost comparison (Table 5.9). It would also have an impact on mask manufacturability (Table 5.8).

A good metric for DfM rule ranking to integrate DRC/DfM checks is the correlation quality between FMEA and RoI. It helps quantifying the impact of layout best practices on the product cost and TTM better than the DRC pass/ fail compliance and can be used to build yield models. Metrics to compare

TABLE 5.9

Cost-Driven Choices between DfM and MfD Responses to Design Issues

Category	Issue	Response	Example	Cost
DfM	Low-frequency bandwidth	Design rules or improved extraction	Larger spaces to dummy fill Fill extraction	IP rework Area penalty
	High leakage	New design rules	Pitch Orientation	IP rework CAD algorithm
	Device mismatch	New OPC and fill engines	Multiple CD control targets	Model calibration Mask complexity
	Frequency shift in time	New design flow	Reliability simulation	Long runtimes Model accuracy
	High power noise	New devices	Fixed layout Wide L of precision Rs	Limited layout options Area penalty
MfD	High leakage	Upgraded fab equipment	New etch for better line profile	New equipment
	Poor yield	New fab process	Exposure correction to tighten CD budget	Process development
	High via R	New materials	New CMP slurry	Cost of slurry Time to etch

designs should include yield as one of the most desired criteria, but they may be difficult to calculate accurately without manufacturing data.

The consequential cost of DfM can also be reduced by reducing the number of standard cells for application-specific and semicustom ICs. Because significant design effort occurs before the first silicon and significant volume manufacturing does not occur until late in the process (as in the 80/20 rule of 80% cost commitment at 20% time into process development), metrics for standard cell area (μm^2), timing or propagation delay (ns), power (dynamic current), load-dependent power, and state-dependent leakage have to be accurate enough to ensure yield improvement for the subsequent products. The SPICE simulations to calibrate the DfM metrics also have to account for manufacturing process variations.

References

1. Aitken, R. DFM metrics for standard cells, *7th International Symposium on Quality Electronic Design*, 491–496, March 2006.
2. Bin, L. Y. New DFM methods enable early yield prediction. *Electronic Engineering Times-Asia*, 9, 1–2, 2006.
3. Ban, Y., Sundareswaran, S., and Pan, D. Z., Total sensitivity based DFM optimization of standard library cells. ISPD, San Francisco, CA, 2010.

4. Raghvendra, S., and Hurat, P. DFM: linking design and manufacturing. *18th International Conference on VLSI Design*, Calcutta, India, 705–708, January 3–7, 2005.
5. Deming, W. E. *Elementary Principles of Statistical Control of Quality*. JUSE, 1950.
6. Venkatraman, R., Castagnetti, R., and Ramesh, S. Modular design of multiport memory bitcells. U.S. patent 7440356, 2008.
7. Fiorante, F., Nicosia, G., Rinaudo, S., and Stracquadanio, G. Automatic layout optimization of power MOSFETs using innovative distributed model technique. *IEEE International Symposium on Industrial Electronics*, ISIE, Cambridge, U.K, June 2008.
8. Kengeri, S., Sabharwal, D., Bhatia, P., Sampigethaya, S., Kainth, S. Multi-port memory utilizing an array of single-port memory cells. U.S. patent 7251186, 2007.
9. Ramesh, S., Castagnetti, R., Venkatraman, R. Memory cell architecture for reduced routing congestion. U.S. patent 6980462, 2005.
10. Unal, N., Mahalu, D., Raslin, O., Ritter, D., Sambale, C., and Hofmann, U. Third dimension of proximity effect correction (PEC). *Microelectronic Engineering*, 87(5–8), 940–942, 2010.
11. Lai, C. M., Liu, R.-G., Shin, I.-C., Ku, Y.-C., and Hou, C. Method for detection and scoring of hot spots in a design layout. U.S. patent application 20070266362, 2007.
12. Harazaki, K. Correction method and correction system for design data or mask data, validation method and validation system for design data or mask data, yield estimation method for semiconductor integrated circuit, method for improving design rule, mask production method, and semiconductor integrated circuit production method. U.S. patent application 20080003510, 2008.
13. Cao, L., and Dahlberg, T. Path cost metrics for multi-hop network routing. *25th IEEE International, Performance, Computing, and Communications Conference, IPCCC*, 22, 10–12, 2006.
14. Huang, W. C., Liu, R. G., Lai, C.-H., Tsai, C. K., Lai, C. W., Tsay, C.-S., Kuo, C. C., and Ku, Y.-C. Regression system and methods for optical proximity correction modeling. U,S. patent application 20070038417.
15. Liu, R.-G., Lai, C.-M., Huang, W.-C., Luo, B., Shin, I.-C., Ku, Y.-C., and Hou, C. Model import for electronic design automation. U.S. patent application 20070265725, 2007.
16. Balasinski, A. Question: DRC or DfM? Answer: FMEA and ROI. Quality Electronic Design, ISQED, 7th International Symposium on 794, March 2006.
17. *FMEA Reference Guide*. http://www.qualitytrainingportal.com/books_consulting/46a_ref.htm.
18. Kahng, A.B. Key directions and a roadmap for electrical design for manufacturability. *33rd European Solid State Circuits Conference, 2007. ESSCIRC*, 83–88, 2007.
19. Yield Enhancement. ITRS Roadmap, 2003.
20. http://www.valuebasedmanagement.net/methods_roi.html.
21. Balasinski, A., Pikus, F., and Bielawski, J. Yield optimization with model based DFM. *Advanced Semiconductor Manufacturing Conference, ASMC, IEEE/SEMI*, 216–220, May 2008.
22. Iandolo, W., et al. Optimization of dummy pattern for mask data size reduction. *Proceedings of SPIE*, 5256, 798–805, 2003.
23. Cetin, J., and Balasinski, A. SoC design quality, cycletime, and yield improvement through DfM. *6th International Workshop on System-on-Chip for Real-Time Applications*, 86–90, December 2006.

6

Summary and Work Ahead

What Have We Learned from This Book?

First, evidence in the form of life examples was provided to show that historical background and recent industry trends are consistent with each other. "New" DfM for IC products and legacy DfM for ME products can be compared. Both are subject to the rule of 10 concurrent engineering, and CBC principles, require RoI analysis with quantifiable metrics such as yield, RPN, or other discipline specific parameters. Because design for manufacturability (DfM) is an extremely multidisciplinary effort of high importance to success of products in the marketplace it is subject to engineering trade-offs in different disciplines, but with the same principles. Setting up a successful DfM approach should be performed throughout an all-inclusive product flow, from inception to end of life. In many mature industries, such as air transportation, medical practice, and civil engineering, proper procedures, when optimized and agreed on, are handled by checklists, splitting the process into multiple phases and making sure that no issue is overlooked.

A checklist approach to DfM acknowledges that all DfM disciplines are equally important when comprehended in the design flow. While the rule of 10 hierarchizes the cost of misstepping at the time of product definition, manufacturing, or testing, it can be equally costly to err in MOSFET (metal oxide semiconductor field effect transistor) material selection for reliability, at the time of pattern transfer to the wafer using a suboptimal optical proximity correction (OPC) option, or to miss test points at the nodes critical to circuit timing.

Secondly, it is easy to realize that this rainbow of DfM issues is typical for any engineering product, and integrated circuits (ICs) are no exception. But, what needs to be rectified—and it is my intention—is that DfM should not be associated only with a narrow domain related to manufacturing, in particular, to the issues of lithographic pattern transfer, at the time of this writing, or with any other narrow domain, depending on future development.

As stated in the Preface, it was not the primary goal of this book simply to recapture the most recent developments in DfM, but to focus instead on the methodological aspects that should better withstand the test of time. The reader might have noticed that the main interest of this look was not in presenting current technical problems in IC DfM and their solutions—existing or stipulative, but mostly reduced to the lithography options for pattern transfer. The narration dealt with finding and exploring DfM patterns that, to me, are more interesting to comprehend and useful to make projections into the future. This is also due to the nature of DfM, which cannot be based on, but welcomes the use of, solid equations and is a discipline of trade-offs and postulates.

Thirdly, the book did not intend to neglect the references to the reality at the time of this writing, and that reality can be synthesized as follows: IC DfM, largely perceived for many years as OPC in disguise, may change its scope with the onset of next-generation lithography (NGL) based on a significantly shorter wavelength of extreme ultraviolet (EUV) (13-nm) light or an equivalent patterning technology. This would partly relieve the pressure from the optical modeling back to the WYSIWYG (what you see is what you get) era, although at the cost of many millions of dollars per mask set. Even then, layout would remain the "final frontier" in the decision on the product yield. In the short term of the next 3 to 4 years, one may expect the DfM agenda to be dominated by the IC architecture conversion issues, from two to three dimensions (2D to 3D), mostly within the ramifications of the existing operating conditions and applications. But with the spectrum of applications growing, new packaging techniques would open up the floodgates to applications in many new product families, from cars to medical equipment to agriculture. CD reduction, by using shorter wavelength, splitting one design layer into multiple precisely aligned masks, or inventing a self-aligned process, may become a trivial issue when compared to the potential need of finding new, highly reliable sensing materials, integrated with power electronics and encapsulated in a fully extractable 3D package. If we take for example, the future of automotive products, the value of electronics would increase from a few to over 50% of their value. Each unit would have its own IP address and an extremely complex—at least by today's standards—diagnostics system. Clearly, all that has to be supported with unparalleled reliability. With printed electronics, expected to be the next "game-changing" technology, the IC penetration into everyday life can increase even further, into large-scale applications such as newsmagazines and maps. It is small wonder that the total IC value was years ago found to be on a trajectory to surpass the total value of all globally manufactured goods—an irrational concept, but only showing the potential of IC DfM—as anticipated for the near and especially for a more distant time frame. But judging by the DfM history, the concepts of the rule of 10, CBC, concurrent engineering, and return-on-investment (RoI) principles would still be valid. It is worthwhile to develop educational programs and bridge the gaps to both user applications and marketing efforts for an even better DfM advantage.

Index